The Volumetric Time Model

Why the Future Feels Decided

Ralph Clayton

Deluxe edition.

Technical Companion

This book is the narrative companion to a formal research program developed by the author in February 2026. The Volumetric Time Model (VTM), Operational Accessibility Theory of Selection (OATS), Agency Horizons, Future Access Without Presence (FAWP), and related constructs are presented here in accessible, story-driven form.The underlying mathematics, information-theoretic definitions, closed-form derivations, Monte Carlo experiments (E1–E8), null controls, bootstrap confidence intervals, and simulation code appear in the following technical preprints (all by Ralph Clayton, freely available on Zenodo)

Preprints (newest to oldest)

- **Forecasting Without Power: Agency Horizons, Control Collapse, and the Leverage Gap**
  Zenodo preprint, uploaded February 19, 2026.
  DOI: https://doi.org/10.5281/zenodo.18700470
- **Future Access Without Presence (FAWP): Operational Prediction Without Leverage in the Volumetric Time Model**
  Zenodo preprint, uploaded February 17, 2026.
  DOI: https://doi.org/10.5281/zenodo.18673949
- **The Agency Horizon: An Information-Theoretic Bound on Action at a Distance**

Zenodo preprint, uploaded February 16, 2026. DOI: https://doi.org/10.5281/zenodo.18663547

- **Future Access Without Presence in the Volumetric Time Model**
  Zenodo preprint, uploaded February 15, 2026. DOI: https://doi.org/10.5281/zenodo.18651290
- **Steering in the Volumetric Time Model**
  Zenodo preprint, uploaded February 14, 2026. DOI: https://doi.org/10.5281/zenodo.18639907
- **The Volumetric Time Model: Constraint-Conditioned Histories and the Operational Origins of Perceived Agency**
  Zenodo preprint, uploaded February 13, 2026. DOI: https://doi.org/10.5281/zenodo.18634216

These preprints contain the exact definitions

Definitions of key quantities, closed form solutions for representative channel classes, and the full experimental pipeline, including estimator choices, autocorrelation preserving nulls, robustness checks, thermodynamic bounds, functional viability boundaries, and quantum consistency constraints. A unified collection page and a maintained code archive for the OATS suite will be provided as project resources. Updates will be noted on the author site or the book resources page. Search for Ralph

Clayton at Zenodo, it will also surface the latest versions and reproducibility artifacts. The narrative chapters translate these operational and geometric ideas into everyday language while preserving the underlying formal commitments. Readers seeking proofs, derivations, or replicable experiments should consult the technical preprints directly.

February 2026

Ducunt volentem fata, nolentem trahunt.
Fate leads the willing and drags the unwilling.
Seneca, Letters to Lucilius, 107.11.

The VTM Model
Future
Past
Present

The Volumetric Time Model

Disclaimer for Skeptics

Everyday Folks, and the People With Too Many Footnotes

This book is for everyday folks. If you have a job, a family, a messy schedule, a brain that gets tired, and a life that does not pause so you can learn new symbols, you are exactly who I wrote this for. You do not need a physics degree. You do not need to be "good at math." You do not need to read slowly like you are taking an exam. You just need curiosity and the honest feeling that sometimes you can see something coming, and still not be able to move the outcome.

That feeling is not weakness. It is not laziness. It is not a moral failure. A lot of the time it is timing. It is access. It is the distance between what you can notice and what you can actually change.

Now, if you are from academia, welcome too. But take a breath. This is not a paper. It is not trying to impress your committee. It is not trying to win an argument on a whiteboard. This book is deliberately written in human language, because normal people deserve the big idea without being punished by the packaging. If you came here hunting for a jargon buffet, you will be disappointed. If you came here to complain that the main text is not drenched in formalism, congratulations, you have correctly identified the design choice.

Here are the common objections, laid out up front, so nobody has to pretend they are discovering them for the first time.

Objection one, this is determinism, so nothing matters.

No. This is not a permission slip for surrender. The Volumetric Time Model is a way to describe a globally consistent reality while taking local limits seriously. You can live inside a consistent structure and still make real choices, because choices are selections among options you can actually reach in time. The point is not that your future is "fixed." The point is that people confuse knowledge with leverage, then feel gaslit by reality when leverage does not show up.

Objection two, this is fate, prophecy, or woo dressed up like physics.

No. This book does not claim you can see the future like an oracle. It claims something narrower and more annoying. Sometimes patterns are visible early, but the lever arrives late. Prediction can persist after control collapses. That is forecasting without power. It is not magic. It is what happens when signals travel faster than usable action.

Objection three, correlation is not causation.

Correct. Thank you. This book agrees. Prediction is correlation with later events. Steering is a measurable influence channel where your action changes outcomes beyond what would happen anyway. These are different. The whole job of this book is to stop mixing them. If you only remember one thing, remember that. Seeing is not steering.

Objection four, you are making philosophical claims from a model.

Some, yes, but not the lazy kind. Models do not dictate morals. Models clarify constraints. The practical claims here are about access and timing, not about cosmic meaning. This is not a sermon about destiny. It is a flashlight pointed at the pipeline between observation and action, and the pipeline can be delayed, filtered, noisy, and broken.

Objection five, this is just metaphors, spheres and fog and cones and corridors.

Yes, because humans think in images, and everyday folks should not need permission to understand. But the claims underneath are operational. You can test them with data and simulations. You can measure when steering becomes indistinguishable from null. Metaphors are not the proof. Metaphors are the bridge, so the proof does not live behind a locked door.

Objection six, people will use this to excuse bad decisions.

They will try. They should not. This framework is not an alibi machine. It is not here to help you dodge consequences. It is here to make one question unavoidable, what could you realistically access, and when, and what levers were actually available. If a system offers you fake choices and late warnings, this book helps you name that. If you had clean access, real time, and a working lever, this book does not protect you from the outcome of ignoring it.

Objection seven, this replaces medicine, law, therapy, or common sense.

No. It does not. This is not medical advice. It is not legal advice. It is not mental health guidance. It is a conceptual framework for thinking more clearly about timing, access, and control. Use professionals for professional problems. Use this book to stop blaming yourself for physics shaped limitations, and to stop trusting systems that only offer you performative buttons.

Objection eight, this is too simplified for experts.

It is simplified for humans. Experts are invited to read the technical companion materials. Everyday folks are invited to read the main text without being shamed. And here is the part academia does not love hearing, confusion is not rigor. Density is not virtue. Jargon is not proof. If your idea cannot be explained clearly, you might not have an idea, you might have a moat.

A friendly note for everyday folks, before we continue.

You do not need permission to get this. You do not need to apologize for asking simple questions. If something affects your life, you are allowed to care about what you can actually change, not what someone can debate forever. You are allowed to say, show me the lever, show me the window, show me the undo button, show me the timing receipt. That is not anti-intellectual.

That is practical, and it is sane.

A closing note for the skeptical reader.

You do not need to agree with everything in this book to get value from it. Treat it like a tool, not a religion. If it helps you separate what you can see from what you can change, and helps you demand real windows, real reversibility, and real levers, then it did its job.

If you are an everyday folk, welcome. This book was written for you.

Prologue

It's Written

You are reading this sentence at a precise coordinate.

Not a vague now. Not a poetic instant. A coordinate in spacetime, a place and a moment along your own path. If you could zoom out far enough, beyond streets and roofs and clouds, beyond the blue curve of Earth, beyond the thin shell of atmosphere that makes life possible, you would not see time as a river flowing past you. You would see a structure.

A finished structure.

Modern physics has been hinting at this for more than a century. Einstein taught us that what one person calls now is not what another person calls now, not if they are moving differently, not if gravity bends their clocks differently. The universe does not hand out a single shared present the way a town clock hands out noon. It hands out local moments, stitched together into a four dimensional geometry.

In that picture, the sentence you are reading is not being written into existence as your eyes move across it. It is already part of the geometry. It sits there the way a mountain ridge sits there, even if you have not climbed it yet. The only thing that is truly happening in real time is your traversal, your worldline, your lived motion through a structure that does not wait for you to arrive.

Most people hear this and feel their mind reach for a familiar story. The story is fate. The story is doom. The story is, if it is already finished, then nothing matters.

That reaction is understandable. It is also too fast.

The point is not whether the volume is complete. The point is what you can reach inside it. The point is what remains locally accessible from where you stand, with the senses you have, the attention you have, the energy you have, the delays you cannot wish away.

Think of the night sky. When you look at a star, you are not seeing the star as it is right now in some universal present. You are seeing an old message, light that left long ago and traveled across space to touch your retina tonight. The star might still be burning, or it might have died and scattered its atoms into darkness. Your view does not change the star. Your view changes you.

This is the first lesson of a finished universe. Access is not the same thing as existence.

Truth can exist far away, and arrive late.

Now shrink the scale from galaxies to your life. The same separation appears, again and again, in smaller and more personal forms.

A storm forms offshore. The forecast is accurate on Monday. The roads are jammed on Tuesday. The water rises on Wednesday. By the time your phone vibrates, the window for the most powerful actions may already be closing. You were told. You were not given leverage.

A medical trend rises silently in a body. The data can see it early. The model can flag it early. The warning can be correct hours in advance. But the appointment is later, the intervention is later, the help arrives after the branch in the road is already behind you. The forecast arrived. The lever arrived late.

A relationship cools in slow motion. The future becomes obvious in the way replies take longer, in the way laughter changes shape, in the way small repairs are postponed. You can predict the ending long before the ending arrives. Yet prediction is not the same as steering. The levers were earlier, smaller, humbler. A conversation before bitterness hardened. A repair before pride took over. A short loop protected before silence grew long.

We live in an age that can see far. We have built instruments that stretch the beam of observation across oceans and continents and bodies and markets. We have built models that can warn, classify, forecast, and rank with astonishing confidence. We have built a civilization that can often describe what will happen before it happens.

And yet, everywhere you look, you see the same eerie pattern. The future becomes clearer at the same time it becomes harder to change.

That pattern is the core of this book.

It is not mysticism. It is not pessimism. It is geometry plus delay.

There is a horizon in time, a distance beyond which your actions stop making measurable differences in outcomes. Before that horizon, the world answers you. After that horizon, the world still unfolds, but your steering fades into background noise. You can push, you can try, you can strain, you can click, you can sign, you can promise, you can plead. Past a certain delay, the effect of those motions collapses.

At the same time, prediction can remain strong. In fact, prediction can become stronger precisely because the future is narrowing. When a corridor has fewer exits, it becomes easier to guess where the crowd will end up. When a complex process approaches a cliff, the path can become more legible because fewer paths remain.

This is why late certainty can feel like punishment.

You see it coming, and you cannot stop it.

You are not powerless because you are ignorant. You are powerless because the lever was not available when it mattered. Or the lever was buried behind friction. Or the lever was presented as a choice that did not actually move the world. Or the lever existed, but the signal was drowned in noise until meaning arrived too late.

That gap, the widening gap between forecasting and steering, is one of the defining experiences of the early twenty first century.

We have become extraordinarily good at producing warnings. We have not become equally good at preserving windows.

A warning that arrives after the last meaningful branch is not a tool. It is a receipt.

A button that records your acknowledgment without giving you a real option is not a choice. It is decoration.

A forecast offered in place of a lever is not empowerment. It is theater.

This book gives a name to the regime where that theater becomes routine. Forecasting without power.

You can think of it as the modern curse of clarity. We can see the wave forming. We can name it. We can chart it. We can quantify it. We can share it. We can even argue about it on screens.

And still, the wave breaks.

The reason is not that the universe is hostile. The reason is that leverage is local. Leverage lives inside short loops. It lives where signals arrive early enough to be usable. It lives where actions can still couple to outcomes. It lives where reversibility exists long enough to reach. It lives where the cost of attention is respected, where meaning is kept legible, where the human body is treated as a real physical device with finite bandwidth, not as an ideal observer that can process infinite prompts at zero cost.

Outside those conditions, control becomes ornamental.

This is why the question that matters is not, is the volume finished. The question that matters is, where is your leverage today.

Where is it in your health. Where is it in your relationships. Where is it in your community. Where is it in your economy. Where is it in your civilization.

In some domains the narrowing is structural. Physics imposes delays. Distance imposes delays. Energy imposes delays. Some doors close quickly no matter how wise you are.

In other domains the narrowing is artificial. Windows shrink because they are made to shrink. Options are delayed because someone benefits when they are delayed. Signals are flooded because confusion is profitable.

Reversibility is withheld because irreversible choices are easier to manage.

The future can be made unreachable on purpose without anyone ever saying the words out loud.

That is why this book is not only about cosmology. It is also about ethics, because ethics is what you get when you translate timing into responsibility. If a person is judged for an outcome, the first honest question is whether a real lever existed within a real window. If the lever was not reachable, then demanding perfect responsibility is miscalibrated. Blame becomes superstition.

This is not an argument for helplessness. It is an argument for precision.

A finished volume does not erase human dignity. It changes where dignity lives. It moves dignity away from fantasies of total control and toward craftsmanship.

Craftsmanship means protecting the loops that still answer. It means making critical signals detectable. It means keeping choices reversible when stakes are high. It means widening effective windows where it is possible to widen them. It means making timing visible. It means refusing to confuse a warning with a lever.

The future will not be saved by better forecasting alone. Better forecasting is easy to celebrate because it feels like intelligence. The deeper achievement is preserving steerability, preserving the conditions where action still matters.

That achievement looks boring. It looks like clear language. It looks like fewer prompts. It looks like undo windows that are long enough to use. It looks like standards that demand that a lever actually moves outcomes, not just interfaces. It looks like simple timing receipts that record when the signal arrived, when the lever was available, and when the door closed.

It looks like a civilization learning the difference between being informed and being empowered.

If you have ever felt the nausea of understanding too late, you already know what this book is about. You have watched a dashboard turn red and then watched the outcome proceed exactly as predicted. You have heard the phrase early warning used like a shield, while the real branch point had already passed. You have felt your own life become legible at the exact moment it became hardest to steer.

This is not a personal failure. It is a geometry that has become common.

The pages ahead will not offer escape from the volume. There is no escape short of leaving physics altogether. What they will offer is a map of leverage, and a way to measure where it dies, and a way to verify that the future can remain readable after influence collapses.

We will build the Volumetric Time Model carefully, not as a metaphor, as a framework that makes this lived experience precise. Then we will test it. Then we will show you what it predicts about responsibility, and why responsibility must track access, not hindsight.

And then, after all the cosmic geometry, we will bring you to a place where the cost of delay is visible in the most undeniable way.

A hospital at 2.13 a.m.

22

Because in a hospital, the words too late are not a feeling. They are a boundary.

Chapter 1

The Warning Arrives Early

At 2:13 a.m., the hospital is a quiet machine pretending to be a place of healing.

The lights never fully dim. They only soften, like someone turned the universe down a notch so the suffering can happen politely. The hallways have that familiar smell—cleaner, plastic, exhausted coffee, the faint metallic note of something medical you can't name but your body recognizes anyway. The kind of air that makes you aware you have lungs.

On the sixth floor, the night shift is running on muscle memory and small acts of denial.

A nurse named Elena—forty, competent, unromantic—checks the vitals outside room 612. Her shoes squeak in the same tired rhythm they've squeaked for years. She's seen every kind of drama, and she no longer believes in drama. She believes in trends. She believes in slow motion. She believes in the way bad news arrives quietly, not with a trumpet.

Inside room 612, a man who looks like a man who has been made to wait his entire life stares at the ceiling. His name is printed on a bracelet nobody reads unless something goes wrong. Let's call him Mr. Kline. He has the damp, pale skin of someone whose body is negotiating with the world in a language the mind doesn't speak.

Elena checks his numbers.

Temperature: slightly up.

Heart rate: slightly up.

Blood pressure: slightly down.

Nothing dramatic. Nothing you can point to in a meeting.

A resident physician—young, sharp, with eyes that still think they can win—scrolls the electronic chart. His badge says "Dr. Patel," but his real identity is caffeine and unfinished sentences. He is covering too many patients, because everyone is covering too many patients, because that is what modern care looks like when you zoom out far enough.

He clicks through tabs like he's flipping channels on a broken television.

Labs. Meds. Notes. Imaging. Orders. Consults.

And then, in a corner of the screen, a red banner flickers into existence.

Sepsis Risk: HIGH.

A number sits beside it. It's not a diagnosis. It's not a verdict. It's a probability dressed like a warning sign. The number is the output of a model that has digested millions of patient records, patterns, subtle correlations that no human can hold in their head at three in the morning.

The model is not emotional.

It doesn't care if Mr. Kline is kind.

It doesn't care if he is alone.

It doesn't care if Dr. Patel is trying his best.

It has one job: recognize the shape of collapse early enough to matter.

Dr. Patel blinks at the banner.

Elena leans in, one hand on the rolling workstation, the other already halfway to the IV pole because she can feel the shift in gravity when a situation changes. She doesn't need a model to tell her when a patient is slipping. But she has learned to respect the quiet things that come before the fall: the slight rise in heart rate, the way someone's breathing becomes a little too shallow, the sweat that appears out of nowhere like a confession.

The model is seeing the same thing she's seeing.

Only it's seeing more.

It's seeing what happens next.

Dr. Patel opens the alert details. There's a list of contributing factors: lactate trending upward, white blood cell count, respiratory rate, recent antibiotics, age, comorbidities. None of them alone are an alarm. Together they form a shape.

A shape the model recognizes.

A shape Elena recognizes, too, but only after she's seen it take someone down fifty times.

Dr. Patel says, "Okay. Let's do the bundle."

Elena nods. She doesn't say finally because she isn't cruel. She doesn't say I told you so because there's no pleasure in being right at a bedside. She just moves.

Fluids. Blood cultures. Broad-spectrum antibiotics. Repeat labs. Monitor closely.

This is the part the public imagines is immediate: the hero doctor hears the warning and charges forward with decisive action, saving the day with science and grit.

Reality is slower.

Reality is a chain.

To start antibiotics, you need an order.

To hang the antibiotics, pharmacy has to verify.

To verify, someone has to check allergies, interactions, dosing.

To check dosing, someone has to confirm kidney function and weight.

To confirm kidney function, you need the lab that isn't back yet.

To get the lab, you need a draw.

To do the draw, you need a person who isn't already drawing blood in three other rooms.

So Elena draws the blood herself.

She has done this before. She will do it again. She ties a tourniquet, swabs the skin, finds a vein that doesn't want to be found. Mr. Kline watches her without speaking, eyes heavy, like he's observing his own life from a distance.

The blood fills the tubes.

Elena labels them carefully. Names and numbers. The ritual of precision.

Then she steps into the hallway and looks for someone—anyone—to take the specimens down to the lab.

The unit is thin tonight. It's always thin. The staffing board is a polite lie. The reality is that nurses and aides have been stretched until "doing your best" becomes a form of triage.

Elena walks the tubes herself.

In the elevator, she stands under fluorescent light that makes everyone look like they're already halfway gone. The tubes are warm in her gloved hand. It's such a small thing to carry: a little liquid, a little data, a little time.

Downstairs, the lab is not asleep. The lab is never asleep. The lab is another quiet machine, chewing through samples, spitting out numbers. But it has its own constraints: how many analyzers, how many techs, how many incoming samples from the entire building. The lab does not care about your urgency. The lab cares about throughput.

Elena hands the tubes across the counter like she's passing a note to the future.

Back upstairs, Dr. Patel places orders. He clicks boxes. He selects "STAT." He adds comments. He documents. He pages. His fingers move fast, but his actions must travel through a maze of checks and queues and approvals that exist for good reasons and bad reasons and "we've always done it this way" reasons.

The warning is early.

But leverage is not.

Pharmacy is backed up.

The rapid response team is in another wing.

The ICU has no open bed.

Radiology is delayed.

The attending is asleep and doesn't love being called unless the numbers look catastrophic.

And here's the cruelest part: the model's warning arrives in the time window where things are still subtle enough that human authority hesitates.

The patient looks sick, but not sick enough.

The numbers are off, but not off enough.

The story is dangerous, but not dangerous enough.

Because humans do not act on the future. Humans act on thresholds.

And thresholds are always late.

At 3:07 a.m., Mr. Kline shivers. A full-body tremor like his cells are trying to shake themselves out of their own skin. Elena notices immediately. She looks at his face, the slight glassiness in the eyes, the way the lips have lost color. She checks his blood pressure again.

Lower.

Dr. Patel sees the updated vitals. He looks back at the red banner. The number is higher now. The model is more confident.

More confident than the humans.

He calls the attending.

The attending answers with a voice that sounds like it was dragged out of sleep by force.

"What's going on?"

Dr. Patel explains. He uses the language young doctors learn early: precise, controlled urgency. Not panic. Never panic. Panic makes people doubt you.

"The sepsis risk model flagged him high," Dr. Patel says, and immediately regrets it, because it sounds like he's outsourcing judgement. He rushes to add, "Clinically he's worsening. He's hypotensive. Tachycardic. Fever. We're starting fluids and cultures. Antibiotics are ordered."

There's a pause, and you can feel, in that pause, the weight of old medicine: the instinct to trust the human eye more than a number.

"Okay," the attending says finally. "Keep me updated."

Not no. Not yes. A deferment disguised as calm.

Elena hangs the first liter of fluid. The IV drips into Mr. Kline like a slow apology. She increases the rate. She watches the monitor. She watches his skin. She listens to his breathing. She watches the clock.

Time does not speed up when you need it.

Time does not care that you were warned.

At 4:22 a.m., pharmacy releases the antibiotic. A runner brings it up. Elena spikes the bag, primes the line, starts the infusion. The medicine enters his body like a promise made late.

By 5:11 a.m., the lactate comes back elevated. The labs confirm what the model was seeing hours ago. Now the numbers look "real" enough for everyone to move. Now the thresholds have been crossed.

Now the building wakes up.

Rapid response arrives with practiced urgency. Respiratory brings oxygen. Someone calls for vasopressors. Someone asks about ICU transfer. Someone asks about code status. People gather at the foot of the bed like a crew arriving to a ship after it has already hit the iceberg.

And Mr. Kline—who has been staring at the ceiling this whole time—turns his head slightly, as if he is trying to locate where all the noise is coming from, as if he is trying to understand why the world suddenly cares.

Elena looks at Dr. Patel. Their eyes meet for a fraction of a second. In that glance is a brutal thought neither of them says out loud:

We knew.

We were told.

We saw the future coming.

And yet the timeline between warning and action was longer than the space in which action could still change the story.

In the morning, when the attending arrives fully awake, he will say something like: "Good catch. You did the right things." And he will mean it. It will be sincere. The paperwork will reflect best effort. The charts will show appropriate escalation.

But the truth will remain: the warning arrived early, and still it arrived too late.

Because being right is not the same as having power.

Because seeing is not the same as steering.

Because there are futures you can predict with high confidence and still be unable to reach with your hands.

You can see it coming.

You cannot touch it.

The Map That Doesn't Care About Your Feelings

By the time the sun comes up, everyone in the building will have an explanation.

The attending will say the patient "declined rapidly."

The chart will say "worsening hypotension."

The notes will say "appropriate escalation of care."

The model will quietly update itself, absorbing the case like food.

People love explanations because explanations are comfort disguised as intelligence. If you can name the thing, you can pretend you had a handle on it. If you can categorize it, you can pretend it was containable.

But the honest story of that night isn't really about medicine.

It's about time.

It's about what the future is allowed to be, and what you are allowed to do about it.

Because look at what happened in that room: the warning arrived early. It wasn't subtle. It was literally written in red. The building was told, in the most modern voice imaginable, "This is where the story goes if you don't intervene."

And still, the building could not turn fast enough.

The future was visible, but the lever was somewhere else—buried under delay, under queues, under rules, under staffing limits, under the soft friction of a world that can predict more than it can change.

This is the shape of our century: knowledge expanding faster than leverage.

We are becoming a civilization of early warnings that arrive too late to matter.

We tell ourselves that time is a river.

It's a good story. It's simple. It has movement. It has drama.

The past is upstream. The future is downstream. You stand in the present like a heroic little boat captain, steering forward into unknown waters. You make choices. You change course. You shape destiny with your will.

It's a story built for the human nervous system, which experiences life as a sequence of "now" moments stitched together like frames in a movie.

But physics has never cared what stories feel comfortable.

About a century ago, Einstein walked into the room and quietly ruined the river.

Relativity takes time—the thing you and I feel as a flow—and treats it like something closer to a coordinate. Not a mysterious stream, but a dimension. Something you can move through in different ways depending on motion, gravity, perspective.

If you've ever heard someone say, "Time slows down near a black hole," or "Clocks tick differently on satellites," you've heard the marketing version of this fact.

The deeper version is stranger: the separation between "past," "present," and "future" is not a universal property of the cosmos. It's a property of how you are embedded inside it.

Different observers can disagree about what is "now."

That sentence should terrify you if you take it seriously. It means that "the present moment" is not a cosmic spotlight shining on one slice of reality. It's a viewpoint—one way of cutting the universe.

And once you accept that, you are only a few steps away from a more unsettling image:

Maybe the universe is less like a river...

...and more like a landscape.

A geography of events.

A completed atlas.

In an atlas, the city you will visit next year already has a street map. The streets are not created by your arrival. They are there. Your arrival simply determines which page you turn to. Which coordinates you occupy.

This metaphor is dangerous because people immediately misunderstand it in the laziest possible way.

They hear "atlas" and think: fatalism.

They think: "So everything is fixed, and I'm just acting out a script."

They think: "So choice is meaningless."

That's not what this book is going to say.

This book is going to make a sharper claim—more useful, more surgical, and honestly more brutal:

Even if reality is globally consistent, you are not globally powerful.

Even if the atlas is complete, you do not hold the atlas.

You are a creature inside the atlas, walking with limited light.

You experience the future as unknown not because it does not exist, but because you cannot access it.

That is the pivot this book is built on.

Not "the future exists."

Not "the future is fixed."

But something far more operational and far more relevant to your actual life:

Existence is not access.

The hospital scene is not a tragedy because the future didn't exist yet.

It's a tragedy because the building had a preview and still couldn't reach the lever in time.

The future was not a fantasy. It was a forecast.

And yet influence collapsed under the weight of delay.

So let's stop telling comforting stories and start describing the machine we live in.

Imagine the universe as a colossal museum.

Not a museum of paintings, but a museum of possible histories—entire lifetimes, entire civilizations, entire sequences of events, arranged not by emotion, but by consistency. Each corridor contains a different chain of cause and effect. Each room is a different version of what could be, constrained by the laws of physics.

Now imagine you are not the curator.

You are a visitor.

You don't get to rearrange exhibits. You don't get to add new wings whenever you feel like it. You don't get to teleport through the building. You walk.

You walk with a flashlight.

The flashlight is your perception. Your memory. Your instruments. Your sensors. Your models. The things you can observe and process.

The beam is narrow.

And here's the cruel detail: the beam has friction.

It doesn't slide instantly through the museum. It moves through walls of delay. Through doors called noise. Through corridors called bandwidth. Through staircases called "processing time." Through guards called "permissions." Through velvet ropes called "rate limits."

The museum is real. The exhibits exist. But your access to them is limited.

That is why you can be surprised.

That is why you can regret.

That is why you can feel like the future "arrived" when, in a deeper sense, you simply reached a corridor you couldn't see earlier.

Now we can say what the Volumetric Time Model is going to do for us.

It is going to stop treating time as a mystical flow that generates reality, and start treating time as a dimension of a consistent structure—while treating your knowledge as a filtered stream, a partial view, a narrow beam.

In other words, VTM is not a religion. It's a way of organizing a problem:

How does an embedded agent—human, animal, machine—operate inside a world that may be globally consistent but locally opaque?

How do you talk about choice without pretending you have god's-eye access?

How do you talk about influence without mistaking "I can predict" for "I can steer"?

Because prediction and steering are not the same thing.

The hospital proved that in flesh.

The model predicted sepsis hours early. Prediction survived.

But the building could not steer the outcome fast enough. Steering collapsed.

That split—between what you can know and what you can change—is the crack in reality where modern life leaks its meaning.

And the reason it leaks is simple: we have upgraded our flashlight faster than we have upgraded our levers.

We have built better sensors than actuators.

Better forecasting than feedback.

Better models than permission structures.

Better visibility than control.

So we live in a strange age where a machine can whisper the future into your ear... while the world politely refuses to move.

If that sounds like a bureaucratic problem, it is. But it's also deeper than bureaucracy, because bureaucracy is just one kind of delay. One kind of friction. One kind of access bottleneck.

Delay is everywhere.

Noise is everywhere.

Constraints are everywhere.

And if you take those seriously, "agency" becomes something you can measure, not something you can argue about at dinner parties.

That's where we're going next.

We are going to build, step by step, a vocabulary for what the hospital scene already taught us:

There is a difference between seeing and steering.

Influence does not stretch forever.

There is a horizon—an edge—beyond which your actions stop leaving detectable fingerprints on outcomes.

But before we get to horizons and the collapse of influence, we need to lock in the first foundational idea of the book. The one that makes everything else inevitable:

The universe may be consistent.

But you are not omniscient.

And you are not omnipotent.

You are a local creature with a limited beam.

So when you ask, "Does the future exist?" you are asking the wrong question.

The right question is uglier and more practical:

What part of the future can you access in time to matter?

Because if you cannot access it, it may as well be behind a wall.

And if you cannot reach it, it may as well be untouchable.

Which is why the most modern form of despair is not ignorance.

It is clarity without control.

In the next part, we're going to sharpen this further. We're going to define what "access" really means in physical terms—delay, bandwidth, noise, energy—and why "knowing the future" is often just another name for reading the present more carefully than the present can act.

Access Is a Physical Thing

When people hear the phrase "existence is not access," they usually nod like it's philosophy.

Like it's a slogan you put on a poster next to a sunset.

But in this book, it's not poetry. It's physics wearing street clothes.

Because *access* isn't a mood. It isn't a belief. It isn't a matter of confidence.

Access is a physical process.

It has speed limits.

It has noise.

It has bottlenecks.

It has an energy bill.

And the strange truth is that we spend most of our lives confusing access with existence—because the human brain is built to assume that what it can't see isn't real yet.

That assumption works for chasing a rabbit. It works for avoiding a fight. It works for day-to-day survival in a world where most of your problems are local and immediate.

It fails spectacularly the moment you step into modern life.

Because modern life is a world of delayed signals.

The hospital is full of delayed signals.

The lab result that arrives three hours later.

The medication that clears pharmacy in forty minutes.

The attending who wakes up at six.

The bed in the ICU that opens tomorrow.

The staffing schedule that was written last week.

Even the warning itself—the red banner—was a delayed signal. A model reading patterns that were already in the patient hours before anyone acted on them.

And that is the point: the future doesn't have to be mystical to be unreachable. It can be unreachable because your access to the relevant information—and your ability to convert that information into action—moves slower than the world.

Think about astronomy for a moment.

When you look up at the night sky, you are not seeing the universe "now." You are seeing light that has been traveling.

Some of it is eight minutes old.

Some of it is thousands of years old.

Some of it is so ancient that the thing emitting it doesn't exist anymore.

You can watch a star and not know it already died.

That is not philosophy. That is delay.

The cosmos is full of events that "exist" in a meaningful sense, but you do not have access to them until their signals arrive—and even then, what you receive is not the event itself, but a degraded trace of it.

A smudge of photons.

A whisper in a detector.

A pattern inferred from noise.

Access is always mediated.

Even in the most intimate situations.

When you tell someone "I love you," you are not transmitting your inner state directly into their skull. You are sending sound waves through air. Their brain reconstructs meaning from vibration. Their history distorts it. Their mood adds noise. Their fear filters it. Their expectations compress it into something manageable.

Your feeling exists.

Their access to it is incomplete.

Every relationship is a channel.

Now take this idea and scale it up.

In the Volumetric Time Model, we treat reality as something like a completed atlas—globally consistent, structured, not being invented moment by moment.

But inside that atlas, what you *experience* is not the atlas. What you experience is an information stream.

A feed.

Not a feed like social media—though that's an excellent metaphor for how degraded access becomes when it gets monetized—but a feed in the physical sense: signals arriving through channels.

And channels have rules.

Some channels are fast.

Some channels are slow.

Some channels are clean.

Some channels are so noisy they might as well be static.

The most important thing to understand early—before we say a single word about agency horizons or forecasting without power—is that access is not a yes/no property.

Access comes in grades.

A spectrum.

At one end: full, immediate, high-fidelity access.

At the other end: faint, delayed, low-confidence access.

Most of what we call "knowledge" lives somewhere in the middle: partial, biased, lagged, expensive.

This is why people fight about free will forever without making progress.

They argue about the metaphysics of choice as if choice is a sacred object floating outside the universe.

But inside a world of channels, the real question becomes much sharper:

How much usable access do you have to the variables that matter before the window closes?

If you can access the relevant information early enough—and you can convert it into action quickly enough—then you experience agency as real. You feel like your choices shape outcomes.

If you can't, you experience reality as a kind of polite inevitability. You watch your life unfold with a strange mixture of prediction and helplessness.

You can be intelligent and still be late.

You can be warned and still be powerless.

This is the modern nightmare: not ignorance, but delayed clarity.

Now we need to make one more distinction that will feel almost insulting in its simplicity, but it is the fulcrum of everything to come:

Access to information is not the same as access to leverage.

The hospital had information access.

The warning arrived.

But leverage didn't arrive with it.

Leverage was trapped behind a different set of channels.

An order channel.

A verification channel.

A staffing channel.

A logistics channel.

A permissions channel.

A bed-availability channel.

Each one with delay. Each one with noise. Each one with bottlenecks.

Information can move at the speed of a click. Leverage often moves at the speed of reality.

And reality is slow.

Here's another thought experiment—simple enough to hold in your head, brutal enough to remember.

Imagine you have a steering wheel in your hands, and you are driving a car.

Now imagine the steering wheel is connected to the tires by a cable that has slack and friction. You turn the wheel and... nothing happens. Then, two seconds later, the tires respond. Not perfectly. A little wobble. A little uncertainty.

If the road is straight, you can still drive. You might even feel confident.

But if the road has a sharp bend ahead, the delay becomes lethal.

You can predict the bend.

You can see it clearly.

But by the time your steering reaches the tires, you are already past the point where steering matters.

That is what delayed leverage feels like.

And it's not rare. It's everywhere.

It is how large organizations behave.

It is how supply chains behave.

It is how health care behaves.

It is how governments behave.

It is how many relationships behave.

The larger the structure, the longer the delay between intention and consequence.

So as our civilization grows more complex—more connected, more automated, more "optimized"—it also becomes a civilization where leverage collapses sooner than prediction.

We are building better telescopes and weaker steering wheels.

We are building better warnings and slower responses.

We are building more accurate ways to see the future and fewer ways to touch it.

This is why the Volumetric Time Model begins not with metaphysics but with access.

Because if reality is a completed atlas, then "the future" is not a magical place. It is a region of the atlas you may or may not have access to.

And access depends on the channels available to you.

Your sensors. Your memory. Your models. Your social connections. Your tools. Your institutions. Your body. All of them are channels. All of them have bandwidth. All of them have delay. All of them degrade. And most importantly: none of them owe you anything.

Physics does not guarantee that meaningful information arrives in time.

Physics does not guarantee that action travels faster than decay.

Physics does not guarantee that being right gives you power.

That's why the hospital warning felt like a cruelty. The model showed the future, but the building's levers were slower than the story.

In the next part, we're going to take this idea—access as a physical thing—and sharpen it into a more uncomfortable claim:

There is a measurable edge beyond which your influence becomes indistinguishable from noise.

Not because you stopped trying.

Not because you lack virtue.

Not because you didn't "manifest hard enough."

Because the channel stopped carrying you.

And once you accept that, you'll be ready for the central twist of this entire book:

A world can remain predictable even after it becomes unsteerable.

And that is where "future access without presence" comes from—no mysticism required.

The Moment Your Fingerprints Disappear

There is a point in every story where your actions stop leaving fingerprints.

Not because you stopped acting.

Not because you became lazy.

Not because you "gave up."

But because the world stopped being able to tell the difference between what you did and what would have happened anyway.

That is the kind of sentence that sounds philosophical until you realize it's literally how physics works.

Drop a pebble into a still pond.

At first the ripples are clean. If someone is watching, they can tell exactly where the pebble hit. They can trace the circles back to a point. They can say, "That's the cause."

But wait.

Let a few minutes pass. Let the wind blow. Let another pebble fall somewhere else. Let the surface fill with overlapping disturbances—tiny waves from insects, distant vibrations from footsteps, the slow breathing of the environment itself.

Eventually, the pond is still again.

Not because the pebble never happened.

Because the information about the pebble's impact has been absorbed into everything else.

The ripple pattern doesn't "travel forever" in a clean way. It smears. It dissipates. It becomes indistinguishable from background texture.

That's the first image you need in your head.

Now replace the pond with reality.

And replace the pebble with your choice.

Most of us carry an intuitive model of agency that goes like this: *If I decide, I cause.* If I choose, the universe tilts. If I want something badly enough, the world should register it.

But the universe is not a diary. It doesn't record your intentions in neat handwriting. It records only physical interactions, and those interactions are mediated by channels.

Channels blur.

Channels delay.

Channels add noise.

So the real question is not "Did you choose?"

The real question is: **Does your choice remain detectable in what happens next?**

If you want an example that doesn't involve ponds and poetry, here's one that every human who's ever yelled at a mountain understands.

Stand at the edge of a canyon and shout your name.

At first, the echo comes back sharp.

You hear yourself.

You recognize the shape of your voice.

It's clearly *you*.

But imagine the canyon is longer. The air is thicker. The walls are irregular. There is wind. There are other people shouting. There are birds. There is traffic somewhere in the distance.

As the echo travels, it gets shredded into pieces.

A syllable disappears.

A consonant becomes mush.

Your voice becomes a smear of sound that could be yours or could be anyone's.

Eventually, if you wait long enough, all you hear is noise.

At that point, you can still shout.

But you can't prove the echo is yours.

Your action has lost its signature.

That is what we mean by a horizon.

Not a mystical boundary in spacetime that forbids magic.

A practical boundary in detectability.

An edge where influence dissolves into the background.

And this is the uncomfortable twist: you can cross that edge while still being fully conscious, fully alive, fully "trying."

The hospital didn't fail because people stopped caring. It failed because the chain from action to outcome was too long, too friction-filled, too delayed. The building could still *do things*, but those things arrived too late to carry the patient back from the slope.

In other words: by the time leverage arrived, leverage no longer mattered.

That is not tragedy as theater. That is tragedy as timing.

Now let's be more precise—but still in human language.

When we say "your fingerprints disappear," we mean that if I look at what happens later, I can't reliably tell whether your action made a difference.

The world becomes a kind of crowded room.

You can speak.

But the room is loud.

Your words may still exist in the air for a moment, vibrating through molecules. But if the room is loud enough, and enough time passes, nobody can tell what you said. They can't even tell you spoke.

In a crowded room, volume matters. Timing matters. Proximity matters.

In a delayed world, *coupling* matters.

Coupling is the strength of the link between what you do and what the world returns to you later.

Strong coupling: turn the steering wheel and the car turns, now.

Weak coupling: turn the wheel and... maybe something happens, later, vaguely.

If coupling weakens with delay—and it usually does—then there will be a time beyond which your influence becomes statistically invisible.

This is the Agency Horizon.

You don't need to love the name. You don't need to frame it as destiny. You don't need to turn it into a lifestyle brand. The only thing you need to understand is that it is measurable in principle, and it changes how you talk about choice.

Because once you accept horizons, the whole free-will debate gets embarrassed.

People argue as if the question is, "Are you free or not free?"

But in a universe of channels, freedom is not binary. It comes in degrees. It is a capacity constrained by delay, noise, energy, bandwidth, and the structure of your environment.

Agency becomes less like a crown and more like a range.

A radius.

A sphere of influence that can shrink or expand depending on conditions.

Inside that sphere, your actions have detectable grip. Outside it, you are still acting, but the world cannot tell.

And here's the truly modern part: our civilization is building realities where this sphere is shrinking faster than our ability to notice it.

Because we are increasing delay everywhere while increasing prediction everywhere.

We have faster sensors, slower levers.

Better warning lights, heavier doors.

More visibility, less reach.

So we end up with a new kind of anxiety that previous centuries didn't have in this exact form: the anxiety of being able to see what's coming while knowing your actions won't register in time.

That anxiety is not irrational. It is not "negative thinking." It is not a personality defect.

It is a rational response to a world where prediction outpaces steering.

Now, a skeptic might say: "How do you know whether influence is detectable? How do you separate 'I caused it' from 'it was going to happen anyway'?"

Good question.

The short answer is: you treat your action like a signal, and the later outcome like a receiver. Then you ask whether knowing the action reduces uncertainty about the outcome.

That's the cleanest operational definition of influence we have.

Not intention. Not blame. Not moral responsibility.

Just: does the world remember what you did?

If the world remembers, you had grip.

If the world forgets, you didn't.

And the world forgets in very specific ways:

- **Delay:** by the time the consequence arrives, other forces have washed over it.
- **Noise:** random variation overwhelms the signature.
- **Rate limits:** you can't act often enough to correct drift.
- **Bottlenecks:** your action must pass through gates and queues.
- **Energy constraints:** you can't maintain tight feedback forever; you tire, you run out, you

degrade.

None of this is mystical.

It is the same reason a radio signal fades with distance and interference. You can broadcast your song, but at some point the receiver hears only static. Not because your song was fake. Because the channel is brutal. So here is the first major claim we're allowed to make, without claiming anything supernatural:

Influence has a horizon.

That horizon is not a philosophical boundary. It is the practical edge where your action-to-outcome coupling becomes indistinguishable from the background.

Past that edge, what do you have?

You still have prediction.

Because prediction doesn't require influence.

Prediction is reading the present.

Prediction is extracting structure from what already exists.

Prediction can persist deep into the future if the present carries traces of what comes next—if the world has inertia, regularity, patterns, constraints.

That's why the sepsis model could see ahead. The patient's body was already moving along a trajectory. The present carried clues. The model read the clues. But steering is different.

Steering requires a return channel that still has grip.

Steering requires that your actions arrive in time, and arrive cleanly enough, to change what would otherwise happen.

Steering is a fight against drift.

And drift is patient. Drift always has time.

So the picture that begins to form is this:

- The universe can be a finished atlas, globally consistent.
- You are embedded in it with a narrow flashlight.
- You can improve the flashlight—better sensors, better models, better forecasts.
- But your levers are separate from your flashlight.
- And the levers have their own delays, their own noise, their own bottlenecks.
- Therefore there will be regions where you can see, but not touch.

That is the region we are driving toward.

That is the region with a name.

But before we name it, we have to do one more piece of surgery: we have to separate two channels that most people confuse as if they are one thing.

The channel that lets you *know*.

And the channel that lets you *change*.

Because once those are separated, the core paradox of modern life becomes not only obvious, but inevitable. In the next part, we'll draw that distinction in plain language and make it unforgettable: There is a channel for forecasting.

And there is a channel for steering.

They can diverge.

And when they do, you enter the strangest state a rational creature can experience:

You are warned early. You are right.

And you are still too late.

Is time a river?

What the hospital taught you, if you're willing to see it, is that there are two completely different highways running through reality. People confuse them because in ordinary life they often travel together. But in modern life they split, and when they split you get that specific kind of clean, cold despair: you can predict the outcome and still be unable to change it. The two highways are these: the channel that lets you *know* and the channel that lets you *steer*. One is about visibility. The other is about grip. One is about information arriving. The other is about influence landing. They are not the same thing, and they degrade at different rates. In room 612, the visibility channel worked beautifully. The warning arrived early. The pattern was recognized. The future left tracks in the present, and a model followed those tracks like a bloodhound. But the grip channel was slow. It had gates. It had queues. It had friction. It had human thresholds that refuse to move until the story looks dramatic enough to justify action. So the building could "see" the bend in the road, while the steering wheel was connected to the tires by a cable with slack. That's the split we're going to live inside for the rest of this book.

Now, it helps to say this in the bluntest possible way: forecasting is not steering. Forecasting is inference. It's reading structure in the present that correlates with what comes next. Steering is intervention. It's sending a signal into the world that meaningfully changes what comes next. Forecasting can get stronger as the world becomes more structured, more measured, more monitored. Steering can get weaker as the world becomes more delayed, more bottlenecked, more rate-limited, more constrained. The terrifying part is that those two trends can happen at the same time. In fact, in our century, they are happening at the same time almost everywhere you look. We are building better eyes and heavier hands. Better sensors and slower actuators. Better models and thicker doors. Better visibility and thinner leverage.

Think of forecasting like astronomy. You don't control a star. You don't push it. You don't negotiate with it. But you can predict it. If you know the mass, the temperature, the chemistry, the rotation, you can estimate its evolution. That's forecasting: the future is not a mystery because the present carries constraints. Now think of steering like piloting. A pilot has leverage because the control surfaces respond quickly enough, and cleanly enough, to stabilize the craft. But if you insert delay—if the plane responds two seconds after the pilot moves the stick—control collapses long before the pilot's knowledge collapses. The pilot can still see the runway. The pilot can still read the instruments. The pilot can still predict the crash. But the pilot can't prevent it, because influence arrives too late to matter. That is the core pattern. A world can be predictable and still be, for you, effectively untouchable.

This is where the Volumetric Time Model becomes more than a metaphor about a finished atlas. The atlas picture explains why the future can be "there" in a meaningful sense without being accessible to you. But the channel split explains something more specific: why your knowledge can outlive your leverage. The world can remain legible even when it stops being steerable. When you treat a person, a hospital, a market, or a civilization as a set of coupled processes moving through time, you can get very good at extracting early warning signals from the present. That's what a risk score is. That's what a trend line is. That's what a model is doing when it says "high risk." It is telling you: given the current state, there is a high-probability corridor you are already walking toward. Forecasting is the art of seeing the corridor. Steering is the art of stepping out of it. And the shocking thing we are learning is that seeing the corridor can be easy while stepping out can be impossible.

Why? Because forecasting and steering are limited by different physics. Forecasting is limited mostly by what information exists in the present and how well you can extract it. Steering is limited by latency and noise in the feedback loop that connects your action to the outcome you care about. In other words, forecasting is limited by the quality of your flashlight; steering is limited by the length and friction of the cable that connects your hand to the world. You can upgrade the flashlight—more data, better instruments, better models—and still be stuck with a cable that is too long to move the tires before the cliff. This is not a tragedy of intelligence. It is a tragedy of coupling.

You can see this in miniature in the hospital. The model doesn't need magic. It doesn't need to "see the future." The body is already sending signals: subtle changes in temperature, heart rate, blood chemistry, breathing. Those signals are present. They exist. Forecasting is reading them as a pattern rather than as isolated facts. The model has been trained to recognize that pattern because it has seen a million versions of it. So it raises the flag. Visibility achieved. But steering requires a different chain: orders, verification, delivery, staffing, attention, authority. Each link adds delay. Each link adds uncertainty. Each link adds a little lost time. And then there is the most human delay of all: the delay of recognition. People hesitate when the story is still quiet. They want proof. They want a threshold. They want drama before they grant permission to act. The model was telling the truth in the low-drama zone, where leverage still existed. The building didn't fully commit until the high-drama zone, where leverage had already shrunk.

This is why we need to talk about horizons. Not cosmic horizons in the astronomy sense, but operational horizons: edges beyond which influence becomes indistinguishable from noise. Because even if you act, even if you push, even if you try, there will be delays beyond which your action no longer leaves a reliable signature on the outcome. And once you accept that, you can stop pretending that "choice" is simply a moral posture. Choice becomes a capacity bounded by timing. In the hospital, the "right choice" was obvious early. The issue was not moral confusion. The issue was that the building's leverage channel could not turn fast enough. You might protest: but antibiotics were given, fluids were hung, teams arrived. Yes. Many actions occurred. The problem is that actions can happen without influence. You can push on a locked door all night and still not open it. The pushing is real. The outcome doesn't change. That is what it means, operationally, for influence to collapse.

Now notice the deeper implication: if forecasting can remain strong while steering collapses, then there exists a strange regime—one that feels like fate even when it is produced by ordinary physics and ordinary constraints—where you can be statistically right about what will happen and still lack the ability to change it. That regime is not a poetic idea. It is testable. It is measurable. It has a signature: prediction remains above the noise floor while influence falls below the detectability floor. In a later chapter we will give this regime a name, and we will show you how to detect it in data without fooling yourself. But for now, in Chapter 1, we only need to lock in the emotional truth: you can be warned early and still be too late. You can know what's coming and still not reach it.

This is why "time as a river" fails as a mental model for the modern world. The river story makes you feel like the present is a magical blade cutting reality into "not yet" and "already happened." It makes you feel like everything downstream is open water you can steer into. But in a world of channels, what matters is not the poetic flow of time; what matters is access and coupling. Some corridors of the atlas are visible long before you reach them, because the present leaks information. But some levers are too delayed, too noisy, too bottlenecked to move the story once it reaches a certain speed. That is why we are surrounded by warnings that feel prophetic and still fail to save us.

So here is the compact thesis of this part, the one you should carry forward like a tool: visibility and leverage are different currencies. You can become rich in visibility and still be bankrupt in leverage. And when that happens, reality starts to feel like a finished atlas not because you are powerless in some cosmic sense, but because your influence has crossed an operational edge. In the next part, we'll put a sharper blade on that idea by turning it into a question that can be answered with data: how do you measure the moment your fingerprints disappear? How do you locate the boundary where influence falls below detection? Because once we can locate that boundary, we can stop arguing about agency like it's a religion and start treating it like something you can map.

If we're going to take this seriously—if we're going to build a book that doesn't just sound profound but actually gives you tools—then we can't leave "fingerprints disappearing" as a metaphor. We need a way to measure it. Not perfectly, not with godlike certainty, but operationally: enough to say, "past this point, influence is no longer distinguishable from background." That's the shift from philosophy to instrument. Because the moment you can measure something, you can stop arguing about it like it's a religion. You can test it. You can falsify it. You can improve it. You can admit when it collapses. And the reason this matters is because the modern world is saturated with arguments about agency that never touch the only thing that counts: whether your actions can be detected in outcomes after delay, under real noise, inside real constraints. In other words: does the world remember what you did in a way that still lets you steer?

So here is the operational trick, stated as plainly as possible. Treat your action like a signal. Treat what you observe later like a receiver. Then ask a simple question: does knowing what action you took reduce uncertainty about what you see later? If yes, then your action left a detectable imprint. If no, then your action might still have "caused" something in some metaphysical story you tell yourself, but it did not create a detectable grip in the only place that matters: the future you are trying to reach. That framing sounds technical because it is technical, but it's also something you already understand intuitively. You've lived it. When you adjust the shower faucet and the water temperature changes immediately, you feel control. When you adjust it and the temperature changes thirty seconds later, and maybe not even in the direction you turned, you feel the ground slide under your feet. It's the same in relationships, markets, organizations, bodies. The delay and noise don't just make things harder. They qualitatively change your experience from steering to watching.

Now we have to say something that will upset the reader who wants a clean heroic narrative: the universe does not care about intention. It does not grade you on effort. It does not file your motives into the official record. The universe only "cares," in the sense that matters here, about physical coupling. You can have noble intentions and still leave no detectable imprint. You can have vile intentions and still leave no detectable imprint. You can press the button and still not change the outcome, not because the button is fake, but because the channel between your press and the outcome has degraded beyond detectability. That's what "horizon" means in this book. It doesn't mean a mystical boundary. It means that beyond a certain delay, the action-to-outcome link becomes so washed out by everything else that, statistically, you cannot separate it from background variation. Your fingerprints don't vanish because reality is cruel. They vanish because noise and delay are efficient at erasing signatures.

This is where people immediately fall into a trap, so we need to flag it early. They say: "But everything is connected. Everything causes everything. So surely my action always affects the future." Yes, in a loose sense. If you clap your hands in a closed room, the air molecules change, and in principle those changes propagate. But that is not the question. The question is not whether some microscopic trace exists somewhere in the universe. The question is whether the trace remains usable for steering. Can you detect it? Can you rely on it? Can you exploit it to change what happens next? If the answer is no, then the trace is philosophically cute and operationally meaningless. A faint ripple that nobody can reconstruct is not leverage. A whisper that arrives after the verdict is not power. This is how you keep the book honest: we do not confuse "influence exists in principle" with "influence exists in practice." The whole point of the Volumetric Time Model, as a framework, is to keep returning to operational accessibility. Not what might be true to a hypothetical omniscient observer, but what is accessible to an embedded agent within real constraints.

Another trap comes from the instant moment—the "right now" coupling. Suppose you take an action and observe something immediately. You might see a strong relationship. You press the pedal; the engine revs. You smile and say, "Look, influence." But instantaneous coupling can be deceptive because it can be shared state rather than steering. You and the world can both be responding to the same underlying condition. Your action and the immediate observation can be correlated because they are both reading the same moment, not because your action is controlling the later outcome. This matters because the modern world is full of dashboards that flatter you with instant feedback while the true delayed consequences are already out of your hands. A button press can make a graph wiggle while nothing meaningful changes in the real pipeline. A policy can create an immediate headline while the underlying dynamics continue unchanged. A person can nod while their future behavior is already locked by constraints you can't see. So when we talk about influence in this book, we will care most about delayed consequences. Not "does something happen right away," but "does your action leave a detectable signature after delay, when the world has had time to wash over it." That is the real test of steering.

So now we can state the first measurable object that will haunt the rest of the book: for each delay into the future, you can ask how much your action helps predict what you observe at that delay. As delay increases, this detectability typically declines, because delay is a machine for destroying signatures. It gives other forces time to intervene. It gives noise time to accumulate. It gives bottlenecks time to randomize outcomes. It gives the world time to overwrite you. And because the detectability declines, there is a point—a delay—where it falls below a chosen threshold of "statistically distinguishable." That is the Agency Horizon. Again: not fate, not metaphysics. A boundary in detectable coupling. Past that boundary, you can still act, but you are acting into fog thick enough that the future cannot reliably tell it was you.

Now notice what this does to the hospital scene retroactively. It doesn't make the staff villains. It doesn't make the model a prophet. It simply reframes the timeline. Early on, there was leverage: fluids could stabilize, antibiotics could blunt the cascade, escalation could change the branch of outcomes the patient was drifting toward. But leverage was trapped behind a delayed action channel. The hospital's steering signals had to travel through approvals, logistics, staffing, threshold hesitation, and physical time. Meanwhile the patient's physiology was traveling forward at full speed. The result is a mismatch: the horizon for meaningful influence was earlier than the moment the building could fully commit. That mismatch is exactly what "being right is not power" looks like when you translate it into channels. The forecast lives in one channel (information). The action lives in another channel (intervention). One can be fast and strong. The other can be slow and weak. And the modern world is systematically arranging itself so that the first channel becomes spectacular while the second quietly collapses.

Here's the cosmic version of that same pattern. A civilization can become extraordinarily good at reading the universe—mapping galaxies, detecting gravitational waves, forecasting climate patterns, predicting economic cascades—because the present carries structure. The universe leaves clues. But the same civilization can become bad at steering its own fate if its levers are slow, fragmented, rate-limited, or captured by delay. It can see the curve and still drift into it. This is not a morality play. It is a control problem. It is the physics of coupling under latency. Which is why the Volumetric Time Model is not merely a philosophical stance about time; it is a framework for diagnosing when and why agency collapses even in the presence of knowledge.

And now we arrive at the final twist we need to set up before Chapter 2: if you separate forecasting from steering, you can finally understand a phenomenon that feels like prophecy but is not prophecy at all. You can have "future access" without "presence." You can have reliable predictive signal about what is coming, even when your influence over what is coming has dropped below detectability. This is the leverage gap. It is the space between what you can know and what you can change. It is where modern anxiety lives. It is also where the most important ethical mistakes happen, because institutions love to punish people as if knowledge implies control. They say, "You should have known," and they treat that as proof you could have acted. The Volumetric Time Model will insist on a harsher, more honest question: did you have operational access to the lever in time, or did you only have access to the warning?

Because a warning is not a lever. A forecast is not a steering wheel.

In the next and final part of Chapter 1, we're going to tighten the whole opening into a clean promise and lay down the formal skeleton in the end-of-chapter Formula Deck. The main text will stay intuitive. But we'll give the reader a small, precise set of definitions at the end: what we mean by a "volume," what we mean by "admissible histories," and how the atlas metaphor maps onto a real mathematical object. Not to intimidate the reader. To prove we're not bluffing. That's the pact of this book: cinematic clarity in the chapter, hard edges in the notes.

A finished atlas is not a prison. It's a reality check. It says: the universe does not have to generate itself one second at a time for you to experience surprise, regret, or choice. Surprise comes from limited access. Regret comes from delayed feedback. Choice comes from coupling that still has grip. In that sense, the most important question you can ask about the future is not whether it "exists," but whether you can reach it in time to change it. The hospital scene makes that question impossible to avoid. The warning was early. The forecast was good. The pattern was visible. But the lever was delayed. The lever had gates, queues, friction, and thresholds. Action traveled through a different channel than knowledge. By the time the building fully moved, the window where movement mattered had already narrowed. That is what modern life increasingly looks like: a world where visibility improves faster than leverage. Better sensors. Better models. Better warnings.

Slower control. Thicker bottlenecks. More delay. More noise.

This is why we began with "existence is not access." Not as a slogan, but as a physical principle dressed in plain language. You can have a future that is structurally predictable because the present contains traces of it. And yet you can be unable to steer because your influence loses its signature under delay and noise. There is a point—an operational edge—where what you do stops leaving detectable fingerprints on what happens later. Past that edge, you can still act, but the world cannot reliably distinguish your action from the background. This is the intuition behind the Agency Horizon: not a mystical boundary, but the practical limit where influence becomes indistinguishable from noise. And once you accept that such horizons exist, the strangest state a rational creature can experience becomes inevitable: you can know what is coming and still be unable to touch it. This is the leverage gap. It is the space where forecasting survives after steering collapses.

The Volumetric Time Model is the framework we will use to map this landscape without turning it into religion. It treats reality as a set of globally consistent histories inside a bounded region—an atlas of admissible stories constrained by physical law. It treats observers—humans, machines, institutions—not as authors rewriting the atlas, but as embedded agents with limited access moving through it with a narrow beam. Your lived "now" is not a cosmic spotlight. It is a local viewpoint carved out by what your channels deliver: perception, measurement, memory, communication. Those channels have delay. Those channels have noise. Those channels have bandwidth limits and energy costs. The world you can reach is the world those channels allow you to couple to in time.

In the chapters ahead, we will build this step by step. First, we will formalize the atlas idea: what it means to talk about a bounded spacetime region, admissible histories, and partial access without slipping into metaphysics. Then we will define agency operationally as detectable delayed coupling, and we will show why delay and noise impose horizons that shrink the radius of influence. Then we will separate forecasting from steering in a way that can be tested in data, and we will define the regime where prediction persists after influence collapses. Finally, we will bring the framework back to the real world: why modern institutions can generate accurate forecasts while remaining structurally unable—or unwilling—to act within the window where action matters, and what that does to responsibility, blame, and design.

If you remember only one thing from this opening chapter, let it be this: the universe is not obligated to be steerable. Being right is not power. Prediction is not permission. Existence is not access.

FORMULA DECK (Chapter 1) The Skeleton Behind the Atlas

1. Spacetime and the bounded region. Let spacetime be represented by a Lorentzian

manifold with metric, written as (M, g). Consider a compact region $V \subset M$ with boundary ∂V. The compactness assumption is used so boundary constraints are well-posed and the region can be treated as a finite "volume of time" for analysis.

2. Fields and admissible histories. Let ϕ denote the set of dynamical variables (fields, states, trajectories) defined on V. Let $E[\phi] = 0$ represent the governing equations on V and let $B[\phi]|\partial V = 0$ represent boundary constraints imposed on ∂V. Define the admissible history set H as the set of all ϕ on V that satisfy both the equations and the boundary constraints: $H = \{ \phi : V \to F \mid E[\phi] = 0$ on V, and $B[\phi]|\partial V = 0$ on $\partial V \}$.
3. The atlas interpretation. Each element ϕ in H corresponds to a globally consistent history within the bounded region V. The Volumetric Time Model treats "what is possible" inside V as this constrained set H. An embedded observer does not possess direct access to ϕ; the observer only receives partial information via channels that deliver data over time. Subsequent chapters will formalize how inference and influence are defined relative to this limited access.

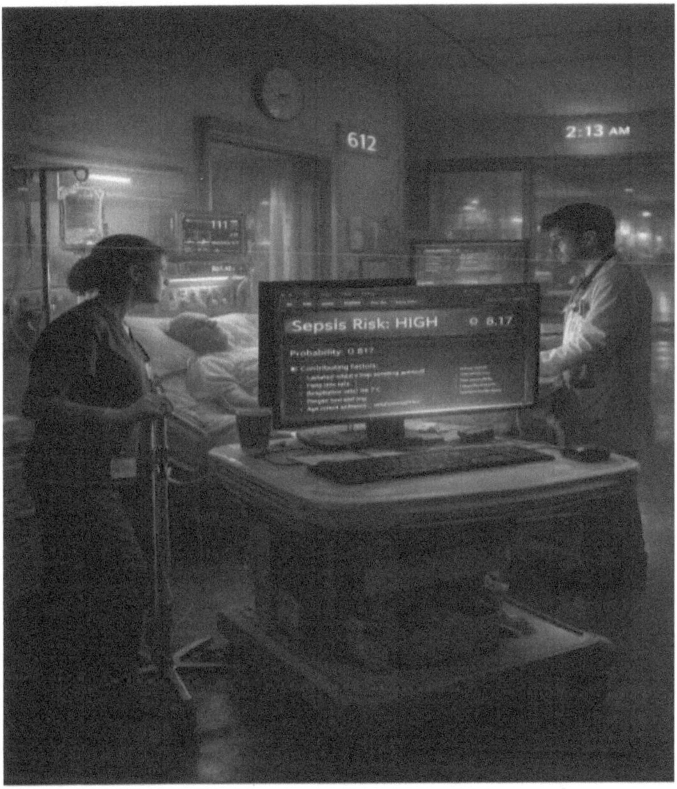
612
2:13 AM
Sepsis Risk: HIGH
Probability: 0.817

Chapter 2

The VTM Model

Imagine you could step outside of time the way an astronaut steps outside a spacecraft, and look back at your life the way you look back at Earth: not as a stream, not as a ticking clock, but as a complete shape. A worldline, etched through spacetime like a glowing thread. From that viewpoint, yesterday and tomorrow are not "gone" and "not yet." They are locations. Coordinates. Parts of one geometric object.

Physics has been pushing us toward this picture for a long time. Relativity already tells you that "now" is not a universal blade slicing the cosmos. Two observers moving relative to each other will disagree about which distant events are simultaneous. If simultaneity depends on motion, then the notion of a single cosmic present is not a law of nature. It is a convenience of human life at low speeds.

The Volumetric Time Model takes the next step and makes that convenience optional. It says: stop treating reality like a conveyor belt manufacturing moments. Treat it like a bounded spacetime volume whose contents are constrained, all at once, by the laws of physics and by boundary conditions. Those constraints do not generate a single inevitable movie. They define an ensemble of admissible movies: complete histories that are allowed because they satisfy the rules everywhere inside the chosen volume.

Start with something concrete. You do not need the entire universe to talk about a real problem. You only ever care about a region: a patient over twelve hours, a storm over two days, a vehicle over ten seconds, a market over a week, a spacecraft during a burn, a brain during a decision. Draw a boundary around the region you care about. Not a metaphorical boundary, a modeling boundary: what counts as "inside" for the purposes of your question, and what counts as "outside." The outside world is not ignored; it appears as boundary conditions. Energy flows in. Signals flow in. Constraints arrive through the boundary.

Inside that region, physics is not negotiable. The equations of motion, conservation laws, and whatever constitutive relations apply act like a set of strict rules. The boundary adds more rules: initial data, final data, fixed values, incoming fluxes, reflective conditions, coupling terms—whatever is appropriate. The result is not a single path. The result is a set of histories that satisfy the whole bundle of constraints. That set is the atlas.

This is the core object: a constrained history ensemble inside a finite spacetime volume. Once you have it, the phrase "the future exists" becomes almost irrelevant, because the interesting question is not metaphysical existence. The interesting question is accessibility. Which parts of that atlas are reachable by an embedded observer? Which parts are inferable from what the observer can measure? Which parts can be influenced by what the observer can do?

That last sentence is not decoration. It is the engine of what makes VTM feel new. It refuses to blur three different layers that people constantly mash into one word like fate. There is a broad field of questions about operational accessibility: how information arrives, how it degrades, how delays and noise and energy limits constrain controllable influence. There is the VTM model itself: the atlas picture, the constrained history ensemble in a finite spacetime volume. And there are derived quantities and regimes—things you compute or detect from within the model—like horizons for influence and conditions where prediction outlives leverage. When you keep these layers distinct, the fog clears. You stop arguing about time like it is a mood, and you start talking about channels, constraints, and measurable couplings.

Now, a fair objection appears immediately: if the atlas is "already there," doesn't that mean everything is fixed? Only if you confuse global consistency with local knowledge. Global consistency means the admissible histories satisfy the rules. It does not mean you, trapped inside the volume, know which history you are in. Your experience of openness is not proof that the world is being created in real time. It is proof that your access is partial.

This is where VTM becomes almost painfully ordinary, and that is its strength. An observer inside the volume does not read a complete history. The observer receives data. Measurements. Signals. Memories. Records. And these do not arrive all at once. They arrive as a growing collection over time, stamped and filtered by the channels that carry them. Your eyes deliver light after a delay. Your instruments deliver readings after processing. Your scans arrive after a pipeline. Your messages arrive after routing.

Your own memory arrives already edited by sleep, emotion, and attention. The cosmos is not waiting. You are the one receiving.

So uncertainty is not a cosmic coin-flip. Uncertainty is bookkeeping for partial access. Early on, many admissible histories remain compatible with the fragmentary data you have. Later, as more data arrives, fewer histories remain compatible. The corridor narrows. You call that "learning." You call it "realizing what was really going on." You might even call it "the moment the truth became real." But the truth did not become real; your access improved.

A simple image helps: imagine you are watching a long eclipse through a telescope, but the telescope feed is choppy. Every few seconds you get a frame. The eclipse does not stutter. Your feed does. At the beginning, with only a handful of frames, multiple trajectories could fit what you've seen. As frames accumulate, the trajectory becomes obvious. Nothing about the celestial mechanics changed at the moment you gained confidence. Your inference changed.

That is VTM's explanation for why the future can feel open while being constrained: you are not observing the atlas, you are conditioning on a partial record. The "present moment" is the edge of that record, not a universal boundary in reality.

If you want to feel how deep this goes, ask yourself why some futures feel inevitable long before they arrive. Why can a skilled clinician glance at a patient and feel dread before the numbers certify it? Why can a meteorologist see the seeds of a storm before the public sees rain? Why can a trader feel a cascade before the chart prints the crash? Because the present contains structure. Constraints make the world legible. The atlas is not random; it is shaped. And if your channels capture enough of the relevant structure, you can infer the likely corridor ahead.

No magic is needed. No information needs to travel backward. You are not peeking into tomorrow. You are reading the present more deeply than someone else, because you have better access, better models, or better intuition about the constraints.

This is also why "existence does not guarantee access" is not a slogan here. It is a physical statement. A future outcome can be strongly implied by constraints and current conditions, and still be inaccessible to you for one of two reasons: you don't have the right data yet, or you cannot convert what you know into timely influence. Those are different failures. The first is an access problem. The second is a leverage problem. VTM keeps them separate so you can measure them separately and stop blaming people for powers they never had.

The model's minimal promise is therefore clean: within a bounded spacetime volume, define the admissible history ensemble by the rules and the boundary constraints; represent the observer as an embedded agent with partial, time-indexed access; represent uncertainty as conditioned weight over admissible histories; and treat prediction and influence as distinct questions about what the observer can infer and what the observer can control before delays and noise erase their signature.

If you hold onto that structure, you gain a new way to talk about modern life that does not collapse into superstition. You can explain why dashboards can be accurate while interventions fail. You can explain why warnings become clear precisely when it is too late to act. You can explain why hindsight feels like moral certainty even though it is often just late-arriving evidence. You can explain why two people can stand in the same reality and inhabit different practical futures because their channels of access are different.

Most importantly, you can make the theory testable, because you've replaced vague metaphysics with operational objects: a constrained history ensemble, a partial data stream, and measurable coupling between what is known, what is done, and what happens later.

To keep the idea sharp, it helps to build a firewall. Not a philosophical firewall—an operational one. Three layers, three different kinds of statements, three different ways people get confused and start accusing you of magic. The first layer is the world itself: whatever is actually happening inside the region you care about, with all its mess, all its microscopic detail, all its hidden variables, all its causal links. The second layer is the model: the way you represent that region as a constrained spacetime volume with an ensemble of allowed histories. The third layer is what you compute from the model and from data: quantities that diagnose what an embedded agent can infer or influence. If you mix these layers, you get nonsense. If you keep them separated, the whole framework becomes clean, testable, and surprisingly intuitive.

Start with the world layer. A real process—say a patient's physiology over twelve hours—is not a set of "moments." It is a continuous physical evolution, with chemical cascades, feedback loops, thresholds, delays, and hidden couplings. The patient is exchanging energy and matter with the environment. The environment is constrained by its own laws. If you could zoom in far enough, you would see a vast web of microstates moving in lockstep with physical law. That world does not care about your vocabulary. It does not contain "past" and "future" as objects. It contains a spacetime pattern of events.

Now the model layer. Here is the move VTM makes that is both bold and ordinary: instead of thinking of the process as something that generates itself frame-by-frame, you treat it as contents of a bounded spacetime region. You choose a region that covers the phenomenon: not the entire universe, just the part relevant to your question. The shape of that region can be anything that makes sense: a twelve-hour slab around a medical episode, a week-long slab around a market event, a two-day slab around a storm system, a three-second slab around a spacecraft maneuver. The reason you do this is not metaphysical. It is practical. If you do not draw a boundary, you never specify what "outside influence" means. You never specify what constraints couple the region to the rest of reality. You end up hand-waving. Bounded regions force honesty.

Once the boundary exists, it plays a starring role. Everything that flows into the region from the outside—energy, signals, interactions—shows up as a boundary condition. Everything that flows out also shows up at the boundary. In physics, boundary conditions are not decoration. They often determine the solution as much as the equations do. In fact, for many problems, equations without boundary conditions are like grammar without sentences. The rules exist, but nothing is specified. The boundary is where the problem becomes a problem you can actually solve.

Inside the region, physical law constrains what can happen. But the key is this: those constraints generally do not pick out a single unique history. They define a set of admissible histories—complete spacetime stories that satisfy the rules everywhere in the region and match the boundary conditions. Think of it as a gallery of allowed films. Each film is internally consistent. Each film obeys the laws. Some films may differ only slightly, in tiny microscopic details. Others may differ in ways that matter at the level you care about: different clinical trajectories, different outcomes, different branches in a decision tree. The gallery is the atlas.

This is where people mishear the idea and think it means "everything is predetermined." But that reaction only makes sense if you assume you, the observer, have access to the entire gallery and can point to the single film that is "the one." You do not. You are inside the region. You are not outside holding the atlas like a god. You are a creature with a flashlight, a notebook, and a noisy stream of delayed measurements. Your lived uncertainty is not proof that the gallery is being drawn in real time. It is proof you don't know which film you are in yet.

That takes us to the third layer: constructs—things you compute, estimate, or diagnose from within the model and from data. This is where VTM becomes more than a reinterpretation of relativity. The model layer gives you an atlas. The construct layer asks: what can an embedded agent actually do with that atlas, given limited access and limited ability to act? This is where later concepts live—horizons for influence, regimes where forecasting persists while control collapses, detection windows where the world is legible but untouchable. Those are not "extra metaphysics." They are measurable consequences of partial access in a constrained world.

The firewall matters because it stops a classic confusion: people see a model that treats futures as part of admissible histories and assume you are claiming an observer can "read the future." But the model does not grant that. The atlas is not your personal library. The atlas is the set of globally consistent possibilities. Access is a separate question. VTM insists that access is mediated by channels, and channels are physical. They have delay. They have noise. They have bandwidth limits. They can be blocked, distorted, or overwhelmed. So the existence of a consistent future corridor in the atlas does not imply any observer can see it, and it certainly does not imply an observer can change it.

A helpful way to picture this without symbols is to think of a sculptor and a block of stone. The stone already contains many possible shapes you could carve out, because the stone has structure and constraints. But the sculptor is not free to carve anything. The sculptor is constrained by the grain, by fractures, by tools, by time, by stamina. The shape that emerges is not created from nothing in a magical present. It is a selection constrained by reality. Now translate this: the atlas is the "many possible shapes" consistent with constraints; your data is the part of the stone you have actually examined; your actions are the tools you actually have; your outcome is the corridor you end up occupying. Nothing about this requires a mystical flow of time. It requires constraints and partial access.

This layered view also clarifies what VTM does not claim. It does not claim messages travel backward. It does not claim you can signal from tomorrow to today. It does not claim your awareness collapses reality into being. Those are different stories. VTM's story is more mundane and therefore more powerful: the present contains information about the future because physical processes have inertia, coupling, and constraints. If you measure enough of the right variables, you can infer which corridors are becoming likely. That's not prophecy. That's physics plus statistics.

The surprising part is how far that inference can reach. In many systems, you can predict breakdown well before it happens. The reason is that failure is rarely spontaneous at the macroscopic level; it is a trajectory. A bridge does not go from "fine" to "collapsed" in one mystical leap. Materials fatigue, stress accumulates, microfractures propagate, tolerances shrink. A body does not go from "healthy" to "shock" in one frame; it drifts through intermediate states with signatures. A market does not crash out of nowhere; fragility accumulates, correlations tighten, liquidity thins, leverage concentrates. In each case, the present leaks the future because constraints tie them together. The atlas is not random noise; it has ridges and valleys carved by physical law.

But again, inference is not the same as control. The ability to say "this corridor is becoming likely" is not the ability to steer into a different corridor. Steering depends on something else entirely: the ability to inject influence into the system in time, through a channel that does not erase your signal. That's construct territory, later. For now, the key is to see how the model creates the stage on which those later questions make sense. Without an atlas of admissible histories, "prediction" sounds mystical. Without partial access, "uncertainty" sounds metaphysical. Without the firewall, everything collapses into sloppy arguments about fate.

One more thought experiment nails the intuition. Imagine a sealed capsule drifting through space, instrumented with sensors, inside a region you decide to study over a finite interval. Outside the capsule, the cosmic environment sets boundary conditions: radiation flux, gravitational field, external signals. Inside, the capsule has its own dynamics. Now imagine you only receive sensor packets intermittently, with gaps. The capsule's actual evolution is continuous, but your record is discrete and delayed. Early on, many internal trajectories are compatible with the few packets you have. As packets accumulate, fewer trajectories remain plausible. You become confident about what the capsule has been doing and what it is likely to do next. At no point did the capsule "become real" when you received a packet. At no point did the future "not exist" until the packet arrived. What existed was a constrained evolution inside the region; what changed was your access to it.

That is VTM in one clean picture: constrained histories in a finite spacetime region; an embedded observer with a partial, delayed record; and uncertainty as the set of still-plausible corridors given what has been observed. This picture is not a substitute for physics. It is a way to keep physics and experience in the same frame without lying. It lets you say, with precision, why the future can be highly inferable from the present without implying any supernatural access. It lets you talk about "openness" as a property of knowledge rather than a property of existence. And it sets up the deeper, more unsettling results later: the possibility that the world can remain legible even after it stops being steerable, because the channel for knowing can stay strong while the channel for influencing collapses.

At this point, the only missing piece is the mechanism by which an embedded observer turns partial data into tightened belief without invoking a mystical flow of time. The mechanism is simple: when new evidence arrives, it shifts weight toward those admissible histories that would have produced evidence like this, and away from those that would not. That is all. No branching universe required. No cosmic dice required. Just constrained possibilities and an observer updating their expectations as access improves.

You can think of your life as a live broadcast, but that's a flattering lie. A live broadcast implies you're receiving the whole scene as it happens. In reality you're receiving a trickle: a handful of photons bounced off surfaces, a few molecules that made it to your nose, a heartbeat you can feel only because nerves are firing, a sentence someone chose to say instead of the ten sentences they swallowed. Even with machines, it's still a trickle. A sensor samples. A pipeline delays. A reading is averaged. A signal is compressed. A screen refreshes. Then your brain does what brains were built to do: it treats the trickle as reality and constructs a coherent world from incomplete data.

That is not a moral failure. That's how an embedded creature survives. But it matters because it explains why the future feels like fog. The fog is not metaphysical. The fog is missing data. The world can be tightly constrained while you remain, locally, blind. You can be standing inside a sharply defined corridor of possible histories and still feel like anything could happen, simply because you haven't received the evidence that would narrow the corridor. The openness you feel is not proof of an unfinished universe. It is proof of an unfinished record.

Astronomy gives a perfect example, because it removes human ego from the equation. When you look at a galaxy a hundred million light-years away, you are not seeing it "now." You are seeing ancient light that has been traveling through space for longer than human civilization has existed. Yet you don't panic and claim the galaxy "doesn't exist yet" until you look. You understand the separation between what is and what arrives. You understand that your access lags behind the object. You understand that your knowledge has a delay. Nobody calls that mystical. They call it physics.

Now bring that back down to Earth. The same separation is happening in every room you've ever been in. You are always receiving delayed information about a world that continues regardless. The only difference is that in daily life the delays are small enough for the illusion of immediacy to hold. The illusion breaks when delays grow, when the environment becomes complex, when measurement is intermittent, when important variables are hidden, when the stakes are high enough that "close enough" becomes "too late." Then you discover the harsh truth: reality does not owe you a clean feed.

So the next piece is simple to state but powerful to feel. At any moment, what you have is a record: everything accessible to you up to that point. Not everything that exists, not everything that matters, just what made it through your channels. This record includes raw perceptions and also instruments, memory, documents, other people's words, images, logs—whatever your world uses to carry information. The record grows. It is cumulative. You don't unsee something once you've seen it. You don't unknow a number once it has been printed. You can deny it, you can reinterpret it, but your accessible record has changed. A new fact entered your world.

Here is where the model becomes almost uncomfortably honest: your belief about "what is going on" is nothing more than a weight distribution over the admissible stories that still fit your record. Early on, with sparse data, many stories fit. Later, with more data, fewer stories fit. When a new measurement arrives that sharply favors one corridor, the weights shift. You call that certainty. You call it clarity. You call it "now I see." But nothing about that implies the universe was undecided before you saw. It implies you were undecided because you did not have access.

The detective metaphor is useful because it feels like life rather than math. A detective walks into a room after an event has occurred. There is one actual sequence of actions that took place, but the detective doesn't get the sequence. The detective gets clues: footprints that could belong to multiple people, a blurry camera angle, a witness who is half wrong, a timestamp that might be off, a broken object that could have been broken in more than one way. At first, several narratives fit. As new evidence arrives—another video clip, a new fingerprint, a confession, a corrected timestamp—most narratives collapse. The detective's uncertainty shrinks. Nobody claims the past was "unwritten" until the detective learned it. The detective is simply converging on which narrative is compatible with the evidence.

The same mechanism is operating in reverse when you talk about the future. You are inside a constrained world. The present carries traces of what is likely to happen next, because physical processes have inertia, coupling, and constraints. You do not have full access to those traces. As you receive more evidence, your beliefs about the future narrow. That narrowing feels like the future becoming "more real." It's not. It's you becoming better calibrated to the corridor you're already in.

A key feature of this viewpoint is that two observers can inhabit the same reality and carry different practical futures simply because their records differ. One person has access to a crucial measurement early; another receives it late. One person sees a trend developing; another sees only noise. One person knows a constraint; another believes it's negotiable. One person has a sensor; another has a guess. This is not mystical; it is a consequence of embeddedness. It also explains a very modern phenomenon: why warnings can exist without changing outcomes. A warning is information. Information changes beliefs. But beliefs do not automatically become leverage. Leverage is a separate question about whether action can couple to outcomes in time. The warning can arrive early and still fail to produce change if the channels for influence are delayed, throttled, or noisy. That separation becomes central later, but it has to be planted here: access and influence are different currencies.

The cosmic version of this is almost poetic. The universe is filled with signals, but not all signals are accessible, and not all accessible signals are useful. The cosmic microwave background carries information about the early universe, but it arrives filtered by the expansion of space and by intervening matter. Gravitational waves carry information about cataclysmic collisions, but they are so faint that they require extraordinary instruments and careful filtering to detect. The events are not created by detection. The detection is the arrival of an accessible trace. In the same way, your daily "events" often feel like they happen when you find out about them. But that's your record talking. The world is not obligated to synchronize with your awareness.

This gives you a cleaner way to think about "choice," without resorting to superstition. People often equate choice with uncertainty: if the future is uncertain, they say, then I am free; if the future is constrained, they say, then I am trapped. But uncertainty can come from ignorance just as easily as it can come from genuine branching at your scale. A roulette wheel is uncertain to you because you don't know the exact microstate, not because physics stopped being law. A storm is uncertain because you don't have full access to initial conditions and because turbulence amplifies tiny differences, not because the atmosphere is writing itself as it goes. Your sense of openness is often a statement about your record, not a statement about reality.

The Volumetric Time Model uses that fact to keep the conversation sober. It says: if you want to talk about openness, talk about access. If you want to talk about inevitability, talk about constraints. If you want to talk about surprise, talk about late-arriving evidence. If you want to talk about guilt, blame, and responsibility, first ask what was accessible at the time and what was not. This is not a trick to erase responsibility; it is a trick to place responsibility where it belongs: on the design of channels, the timing of access, and the practical ability to act within the window where action still leaves a signature.

A subtle but important point follows. The record you carry is not just partial; it is structured by what your channels consider "legible." Some facts arrive with authority. Others arrive as rumor. Some arrive with a number and a timestamp and a signature. Others arrive as intuition. People treat the authoritative facts as "real" because they are defensible. That's why a team moves differently when a certain number prints. Not because the body changed at that moment, but because the social weight of the record changed. The corridor became harder to deny. Your beliefs are not updated by raw reality; they are updated by what reality manages to deliver in a form you can use.

This is why the model insists on separating the world from the record. The world is what it is. The record is what made it through. The record is what you condition on. If you forget this, you will keep making the same mistake: you will treat late-arriving evidence as late-arriving reality. You will treat surprise as creation. You will treat uncertainty as proof the future is unwritten. And you will keep being emotionally ambushed by outcomes that were constrained long before they became visible to you.

The next step is to make the bounded volume idea feel less like a poetic metaphor and more like a disciplined move you can apply anywhere. What does it mean to draw a boundary around an event? What counts as a boundary constraint in real situations? Why is the "volume" not a claim about the size of the cosmos, but a way of turning a messy world into a well-posed question? Because once you have that, the atlas stops being a story and becomes a tool: a way to say, with clarity, what is allowed, what is unknown, what is accessible, and what is not.

A boundary is not a wall. A boundary is a promise. It's you saying, with discipline, "This is the region I'm talking about, and everything outside it will be treated as a constraint rather than a mystery." Without that promise, you're not doing physics, you're doing poetry. You're waving your hands at "the whole universe" and pretending it explains your specific event. In real science you always carve out a region, even if you don't admit it. You carve it out with instruments, with time windows, with what you choose to measure, with what you call relevant. The Volumetric Time Model simply makes that carving explicit and puts it at the center: pick a finite slice of spacetime that contains the phenomenon you care about, then let the laws and boundary conditions define what histories are allowed inside that slice.

The reason this matters is that the universe is not a collection of isolated objects. Everything exchanges something with its surroundings: energy, matter, momentum, radiation, signals. Even a "closed" experiment is only closed by approximation. So when you draw a boundary around your region, you are not pretending the outside doesn't exist. You are acknowledging it in the only honest way available: by describing how the outside couples to the inside. That coupling is what boundary conditions are. They are the handshake between your chosen volume and the rest of reality.

Think of a soap bubble floating in air. The bubble is not sealed off from the world. Air pressure presses on it. Temperature gradients stretch it. Air currents deform it. The bubble has a surface that mediates what passes in and out. Your chosen spacetime volume is like that bubble. The boundary is the surface. The "outside" influences the inside through what crosses that surface. You might specify that certain quantities are fixed on the boundary, or that certain fluxes enter, or that certain values match an environment. You might specify initial information at one side of the volume and final information at another. You might specify both. The point is: once you define the coupling, the inside is no longer free to do anything it wants. It must satisfy the equations everywhere inside, and it must match the boundary everywhere on the surface.

This is where the atlas comes from. The atlas is not imagination. It is the set of all spacetime histories inside the bubble that obey the rules and fit the boundary coupling. Each admissible history is a full four-dimensional story within that volume—what you would call the earlier portion, the later portion, and everything in between. These are not "moments." They are consistent patterns in spacetime. The reason they feel like "movies" is because your brain thinks in sequences, but the object is geometric: a complete configuration spread across the volume.

In everyday life, people talk as if the past is fixed because it already happened, and the future is open because it hasn't happened yet. That language is psychological, not physical. Physically, the constrained volume doesn't care which part you label "past" and which part you label "future." The constraints apply everywhere. If you pin down enough information on the boundary, you can narrow the admissible set dramatically. If you pin down less, the admissible set is broad. That breadth is not randomness inside nature; it is under-specification from your viewpoint. It is the difference between a tightly constrained problem and a loosely constrained one.

A simple example: imagine a guitar string. The wave equation tells you what oscillations are allowed. But the actual vibration depends on how you hold the ends and how you pluck it. Fix the ends and choose the initial shape, and the evolution is determined within that idealization. Change the boundary—loosen one end, damp it, attach it to a heavier object—and the admissible vibrations change. The laws didn't change. The boundary coupling changed. The family of allowed histories changed. That's the theme: the rules set the grammar; the boundary conditions write the sentence.

Now scale that up to something that feels like life. A patient in a hospital bed is not an isolated object. The body exchanges heat with the room, oxygen with the air, fluids with the intravenous line, chemicals with medications, information with monitors and clinicians. If you draw a spacetime volume around a twelve-hour episode, the boundary includes what enters and leaves: interventions, environmental conditions, initial state, and the constraints of biology. That boundary coupling matters because it shapes what trajectories are admissible. It determines which spirals toward collapse are compatible and which recoveries are compatible. If you change the boundary coupling—deliver oxygen earlier, correct fluids earlier, remove a toxic agent—then the admissible set shifts. Again: not because time was created differently, but because constraints were different.

Or take a storm. The atmosphere is a fluid with equations that do not bend for your feelings. But the storm's trajectory depends on boundary conditions: ocean temperature, pressure fields, jet streams, humidity gradients, terrain. Draw a spacetime volume around a region for two days and specify the coupling at the boundaries, and the admissible set becomes a family of possible storm histories. Some corridors lead to dissipation; others lead to intensification. The present contains hints about which corridor is more likely, because constraints create structure. But the future is not "unreal" until it arrives. It is unknown to you until enough evidence is accessible.

This is where a common misunderstanding creeps in. People hear "bounded volume" and think it implies you're claiming the universe is literally a finite box. That's not necessary. The bounded volume is a modeling lens, like choosing a coordinate chart. You're not claiming the rest of reality doesn't exist; you're choosing a region where the question is well-defined. Every scientific calculation does this. When an engineer analyzes the stress on a bridge beam, they do not model the entire planet. They model the relevant structure and represent the environment as boundary loads and constraints. When a physicist studies a collision, they choose a region around the interaction and treat the outside as asymptotic conditions. When a neuroscientist studies a circuit, they choose a timescale and treat the rest as inputs and noise. The Volumetric Time Model just makes this habit explicit and then uses it as the foundation for thinking about time, access, and agency.

Another subtlety: boundaries don't have to be literal walls. They can be practical. Your instruments impose boundaries. Your sampling rate imposes boundaries. Your choice of variables imposes boundaries. If you measure a body once per hour, you have implicitly chosen a coarse time boundary; you have declared everything that happens inside those hours as invisible detail. That invisibility doesn't erase the detail from the world, but it changes what's accessible, and therefore it changes what you can infer. The record you carry is shaped by these practical boundaries. In that sense, "the volume" is not just an abstract region of spacetime; it is the region your sensing and logging actually cover.

This is why the model has teeth: it treats access limitations as first-class physics, not as an embarrassment to be ignored. Most human frustration comes from confusing the world with the record. You act as if the record is the world because it's all you have. Then you are shocked when the world behaves as if it knew more than you did—because it did. It contained hidden variables, unseen couplings, delayed effects, constraints you didn't measure. The model reframes shock as what happens when your partial record fails to pin down the corridor you're actually in.

There is also a deeper reason boundaries matter, and it touches the ancient human question of why time feels like it "flows" at all. In a constrained volume, a complete history is a geometric object. There is no intrinsic "flow" in the object itself. The sense of flow arises for an embedded observer because the observer receives information in sequence and stores it as a record. The record grows. The observer updates. Memory accumulates. Evidence arrives. That sequential arrival creates a psychological arrow: a direction from less information to more information, from many plausible corridors to fewer plausible corridors. That is why the present feels like a blade even though physics doesn't require a universal blade. The blade is the edge of your record.

Now notice something important: even if the underlying physics is perfectly deterministic in the idealized sense, an embedded observer can still live inside uncertainty because they don't have perfect access to initial conditions, boundary coupling, or hidden variables. Determinism is a statement about the rules, not a guarantee of omniscience. You can have a fully constrained world and still have an observer who cannot recover which corridor they occupy because the record is incomplete or noisy. This is the key to avoiding fatalism: global consistency does not imply local certainty.

So when you hear "admissible histories," don't imagine a cosmic librarian stamping each history as "real." Imagine instead a family of spacetime patterns that satisfy the constraints. The job of an observer is not to invent a future. The job of an observer is to infer, from partial access, which subset of patterns remains compatible with what they've seen. As their record grows, that subset shrinks. That shrinking is what you experience as learning, as clarity, as certainty. The world did not become more real. Your access became less blind.

And because boundaries are where coupling lives, they are also where intervention lives. If you change what crosses the boundary—change the inputs, change the constraints, change the coupling—you change the admissible set. This is how VTM makes room for action without pretending action rewrites reality from outside. Action is not supernatural. It is just a physical change in boundary coupling or internal dynamics, performed by an agent who is themselves inside the volume, with limited access and limited ability to affect the relevant variables in time. Whether that action actually produces a detectable difference later is a separate question about delayed coupling, but the stage on which the question is asked is the bounded volume and its admissible histories.

A bounded spacetime region is therefore a lens that turns vague talk into a well-posed question: what histories are consistent, what evidence is accessible, what corridors remain plausible, and what couplings can still be altered before the world's constraints make the outcome effectively locked from the observer's viewpoint.

If you want a single sentence that captures how an embedded mind survives inside a constrained universe, it's this: you never hold the whole story, you hold a growing stack of clues. Each clue arrives through a physical channel. Each channel has delay, noise, distortion, and blind spots. And because your stack of clues grows in one direction, you experience that direction as time "flowing." The feeling is real. The cause is local. The cosmos does not need a universal spotlight called NOW for you to feel the arrow of experience. The arrow is the accumulation of accessible evidence.

Look up at the night sky and you can feel this separation with almost comic clarity. The stars are not sending you their "present." They are sending you ancient light. Your retina is a museum. Even the Sun, that blazing certainty, is already eight minutes old by the time it reaches you. Yet nobody panics about whether the Sun "exists" while you wait for photons. You accept the obvious: there is a difference between what is and what arrives. You accept the delay as part of the world. Now take that same honesty and bring it back to Earth, into your own day. Every meaningful fact you learn arrives the same way: as late evidence, sampled and filtered. You don't experience raw reality; you experience what reality manages to deliver in usable form.

The moment you accept that, probability stops being philosophical fog and becomes simple accounting. Imagine a locked room containing a complex machine. You cannot open the room. You only receive a weak stream of sensor readings that occasionally glitch. You want to know what the machine is doing. There is one actual physical evolution happening inside that room, but you don't have it. You have partial access. Early on, many internal trajectories are consistent with the sparse readings you've seen. Later, as more readings arrive, fewer trajectories remain consistent. Your uncertainty shrinks. This shrinking is not the machine creating itself in response to your attention. It is you eliminating possibilities that no longer fit the evidence. That is all probability is in this framework: weight assigned to still-viable histories given a partial record.

This is why the "unfinished future" story is seductive but unnecessary. It explains openness by inventing a cosmic factory that manufactures reality frame by frame. But you don't need a factory to explain openness. Openness is what incomplete access feels like from the inside. The world can be sharply constrained while you remain locally uncertain because your record is incomplete. Your beliefs are broad because your evidence is narrow. When new evidence arrives, your beliefs narrow. The corridor becomes visible. And the psychological punchline is brutal: when the corridor becomes visible, it often feels like it was "obvious all along." It wasn't obvious. You didn't have the clue yet. You are mistaking late access for late reality.

The word "conditioning" is just the clean name for this process of narrowing. You start with an initial spread of plausibilities over what could be happening inside the bounded region you care about. Then you receive data. You tilt your plausibility toward the histories that would naturally produce data like this and away from the histories that would not. If the data is weak, the tilt is gentle. If the data is strong—an unmistakable signature—the tilt is violent. It's like tuning a radio. At first you have static and multiple stations bleeding into each other. As you turn the dial and refine your reception, one station dominates. The broadcast didn't change when you tuned it. Your access improved. Your belief locked onto the correct channel.

Once you see this, you can explain something that feels almost supernatural in daily life: the same event can be "real" for one person and "not real" for another until an evidence packet arrives. One person sees the warning signs early because they have access to the right measurements or because they have a model that knows which faint signals matter. Another person doesn't see it because their access is delayed or their signals are drowned in noise. Then the decisive clue arrives, and everyone acts as if reality changed at that moment. But nothing about the underlying process needed to change. The corridor was already tightening. The difference was access.

This is also why a bounded region matters so much. A bounded region turns "everything influences everything" into a question you can actually ask. Once you draw the region, you can list what enters through the boundary and what leaves. You can admit what you measure and what you don't. You can admit what you treat as known and what you treat as unknown. Then your belief updating becomes disciplined rather than mystical. The admissible histories inside the region are those that satisfy the rules and match boundary coupling. Your accessible record is what your channels captured from inside that region. Your uncertainty is the set of histories still compatible with that record. The universe doesn't need to roll dice. You just need to admit what you didn't observe.

There is a more profound consequence that begins to appear here, one that will later become central: knowledge is a physical object with a cost. It takes energy to measure. It takes energy to store. It takes energy to communicate. It takes time for signals to propagate. It takes time for instruments to process. Even in the cleanest physical picture, access is limited by fundamental constraints. Information cannot arrive instantly from far away. High-fidelity measurement cannot be free. Perfect records do not exist for embedded creatures. This means the "flow" of experience is anchored in physics, not in metaphysical storytelling. You experience a moving boundary between what you have and what you don't have because you are made of matter, and matter obeys limits.

Now consider what happens when you apply this lens to the most emotionally charged human words: certainty, surprise, inevitability. Certainty is not an intrinsic feature of the world; it is the narrowness of the set of histories consistent with your record. Surprise is not the universe changing its mind; it is the moment your record receives a clue that kills the corridor you were leaning on. Inevitability is not fate descending from heaven; it is what it feels like when the admissible set has a dominant corridor and your record has already narrowed onto it. You can still be wrong—records can be misleading—but the sensation is explainable: your access has already collapsed your belief.

Here is the paradox that makes this more than just a clean story: the present can contain enough structure to allow confident prediction while still leaving you powerless to change the outcome. That sounds like a contradiction only if you assume knowing and influencing are the same. They are not. Knowing comes from evidence flowing into you. Influencing comes from your actions coupling back into the world in time. Those are two different directions of flow, two different physical channels, with different delays, different noise, different bottlenecks. A civilization can become brilliant at extracting structure from the present while becoming clumsy at injecting timely influence. A person can see the cliff and still drift toward it. The model does not need prophecy to explain this; it needs only the separation between access and leverage.

But before that separation can be measured, one more piece has to be installed: the idea that your belief is not about single points in time, but about whole histories. When you update on new evidence, you are not merely changing your opinion about "what is happening right now." You are changing your belief about the entire pattern inside the bounded region: where you came from, what is happening, and what comes next. New data reaches backward in your interpretation because it helps eliminate histories that would have produced a different past pattern. This is why late evidence can make you reinterpret earlier events. It doesn't change the earlier events. It changes which histories remain compatible with all the evidence. In an atlas picture, one new landmark can tell you that you weren't on the road you thought you were on five miles ago. The road didn't move. Your map match improved.

In that sense, the model is almost mercilessly anti-mystical. It refuses to grant the observer magical privileges. You don't get to stand outside the atlas. You don't get to know the whole story. You don't get to declare the future "unreal" just because you haven't received the clue yet. What you get is exactly what physics allows: a partial stream of evidence, arriving with delay, which you use to assign weight across admissible histories. As evidence accumulates, your belief narrows. The arrow you feel is the arrow from less access to more access. And once you accept that, you can finally ask the next practical question without lying to yourself: if your beliefs can narrow onto an outcome long before it arrives, what determines whether you can still push the world onto a different corridor before the window closes?

If the universe were a cathedral, you would not live on the roof where the whole architecture is visible. You live inside one hallway with a flashlight and a pulse. You hear echoes. You feel temperature shifts. You catch fragments: a distant rumble, a faint draft, a sliver of stained glass. From those fragments you build a story about the whole cathedral, and you call that story "the world." Most days the story is good enough. The flashlight is bright enough. The hallway is familiar. Then something big happens—an approaching collapse, a sudden cascade, a turning point you swear came out of nowhere—and you discover what you were actually living in: partial access. Not ignorance as a personal flaw, but partial access as a physical condition.

That condition is universal. Every observer is embedded. Every observation arrives through channels. Every channel has limits. Even the cleanest instrument is a translator, turning the world into a legible record by sampling, averaging, compressing, and delaying. The record grows. It grows one way. That growth is the arrow you feel. It is the only arrow you ever directly touch. And because your record is incomplete, the set of plausible stories about what is happening remains broad until enough evidence arrives to narrow it.

Now place that embedded observer into a bounded spacetime volume—an honest, finite region containing the phenomenon you care about—and the whole picture becomes crisp. Within the region, physical law and boundary coupling carve out a family of allowed histories:

complete spacetime patterns that obey the rules everywhere inside. That family is the atlas. The atlas is not a list of fantasies. It is what remains after constraints have done their work. It has structure. It has ridges and basins. It has narrow corridors that many trajectories funnel into, and wide plains where outcomes spread. The atlas is a physical object in the sense that it is defined by physical law and boundary constraints, not by imagination.

The embedded observer does not "see the atlas." The observer sees a stream of evidence. The stream is sparse compared to the underlying reality, and it is delivered with delay. So the observer does not hold a single story; the observer holds a distribution of plausibilities over which histories in the atlas remain compatible with the record so far. This is why probability belongs here without metaphysics. It is not a cosmic coin-flip. It is your accounting method for partial access. When new evidence arrives, you shift weight toward those admissible histories that would naturally produce evidence like this, and away from those that would not. The atlas doesn't redraw itself. Your belief over the atlas tightens.

That tightening, when you feel it from the inside, is what people call reality "becoming real." A number posts. A scan resolves. A message arrives. A sensor finally captures the missing signature. Suddenly the corridor you were already in becomes undeniable. The world didn't change because you learned. You learned because the world leaked a trace into your record. You updated because access improved. If you keep that distinction—world versus record—you stop making the most common mistake humans make when they talk about time: treating late-arriving evidence as late-arriving existence.

The three-layer firewall matters because it prevents a second, equally common mistake: confusing the atlas with supernatural foresight. The atlas is a model object: the family of consistent histories inside a bounded region given constraints. Your record is an access object: the data that made it through your channels. Quantities you compute from these—diagnostics of inferability and influence—are construct objects. When people accuse you of claiming time travel, they are usually smearing these layers together. They imagine that if complete histories exist in the atlas, then an observer must be able to read them. That does not follow. Existence does not guarantee access. The atlas can be rich while your record remains thin. The future can be tightly constrained while you remain locally uncertain. All that is required is that you are embedded and your channels are limited.

This is also where the most unsettling modern experience becomes explainable without drama: the present can be legible while still being hard to steer. Constraints give the world structure. Structure makes inference possible. Your knowledge channel can remain strong because the present carries signatures of what is likely next. But your influence channel can degrade because your actions arrive late, or arrive too weakly, or are absorbed by background variation. In other words, you can see the cliff in high definition and still be unable to turn the wheel. That mismatch is not fate. It is architecture:

different channels with different delays, different noise, different bottlenecks.

Once you understand that, a new kind of scientific humility appears. You stop asking the universe for a privileged present. You stop demanding that the future be "unreal" until you personally arrive. You accept that your lived "now" is the edge of your record. You accept that your sense of openness is a property of what you can access. You accept that certainty is a property of how narrow the compatible corridor has become given the evidence. You accept that surprise is often just late data killing the story you were leaning on. And you accept that responsibility, in any serious sense, must be discussed in terms of what was accessible and what couplings were available within the time window that mattered. Not as an excuse, but as an audit of reality as experienced by embedded agents.

In this view, the universe is not a movie being projected onto a screen called the present. It is a constrained four-dimensional geometry, and you are a creature moving through it with a notebook that fills up one line at a time. The notebook is not the world. It is what the world has delivered to you so far. Your predictions are not prophecies. They are inferences conditioned on partial access to a structured atlas. Your actions are not magical edits of reality. They are physical perturbations whose ability to matter depends on whether their signature survives delay and noise long enough to alter which corridors remain reachable.

There is a quiet grandeur in that picture. It makes time less mystical and more physical. It replaces sentimental metaphors with measurable objects: bounded regions, constraints, admissible histories, accessible records, conditioned belief. It respects relativity's lesson that "now" is not a universal law, while still explaining why "now" feels like a knife-edge to a human nervous system: because evidence accumulates, memory stacks, access arrives in sequence, and your mind uses that sequence as the backbone of experience.

And it gives you a way to talk about the future that is both sober and strangely liberating. The future does not need to be "unwritten" for you to have uncertainty. Uncertainty is what it feels like to stand inside a vast atlas with a small flashlight. You can still make choices, but those choices are constrained steering among reachable corridors, not wishful selection among fantasies. You can still be surprised, but surprise is often a correction of belief by late evidence, not the universe improvising. You can still be warned early, but a warning is only power if the channel for influence remains above threshold. The world can be coherent while you remain partially blind. That is not a defect in reality. It is the condition of being a finite agent inside a finite volume of time.

FORMULA DECK (Chapter 2)

Access, Filtration, and Conditioning Over Histories

1. Observer data stream. Let spacetime be represented as (M, g). Fix a bounded region V

inside M with boundary ∂V, and let H be the admissible history set on V as defined previously. An embedded observer does not access a full history directly. Instead, the observer receives a growing record of accessible data over time. Call that record D_t, meaning the data accessible up to time t.

2. Information set as a filtration. The observers available information so far is represented as an increasing family of information sets generated by the record. Let F_t denote the information generated by D_t. As time increases, the accessible record grows, so access is monotone. Later information includes earlier information.
3. Prior weighting over admissible histories. Before conditioning on the specific record, the observer has a baseline weighting over admissible histories in H. This prior weighting encodes what histories are considered more or less plausible before data arrives.
4. Conditioning on the record. Given the accessible record D_t, the observer updates from the prior to a posterior distribution over histories that remain compatible with what has been observed. Histories that would be unlikely to produce the observed record receive lower posterior weight. Histories that would likely produce the record receive higher weight.
5. Normalization. The posterior is normalized so that total probability across H sums to one. This

normalization step does not add new information; it only ensures the posterior is a valid distribution.

6. Posterior predictive over future observables. Let a future observable mean any later measurement or outcome defined inside V at a nonzero delay. The posterior predictive distribution is obtained by averaging predictions across admissible histories, weighted by their posterior probabilities. This is the formal sense in which the observer can forecast the future using partial access.
7. Interpretation rule. This formalism separates global consistency from local access. The admissible set H is fixed by constraints on V. Uncertainty evolves because the accessible record grows and the observer conditions on it over time. Openness is modeled as partial access and belief updating, not as creation of new histories.
8. Bridge to later chapters. Conditioning can improve forecasts even when steering collapses. Later chapters will show how prediction and influence separate under delay and noise, which is the doorway into agency horizons and the Forecasting Without Power regime.

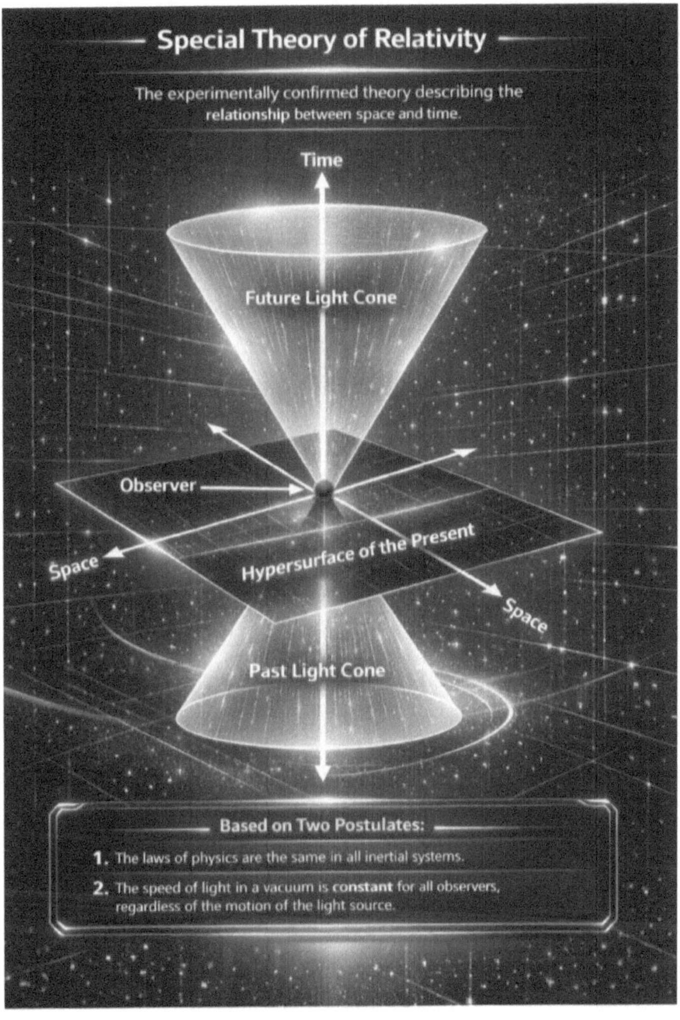
Special Theory of Relativity
The experimentally confirmed theory describing the relationship between space and time.
Time
Future Light Cone
Observer
Space
Hypersurface of the Present
Space
Past Light Cone
Based on Two Postulates:
1. The laws of physics are the same in all inertial systems.
2. The speed of light in a vacuum is constant for all observers, regardless of the motion of the light source.

Chapter 3

The Observer Is a Data Pipeline

Light that left a dying star before humans invented writing can still fall onto your retina tonight. That single fact should permanently cure you of the fantasy that you live in "the present." You live in delayed evidence. You live in arrivals. The cosmos is constantly mailing you postcards from events that are already finished, and your brain—this exquisite, impatient machine—calls the latest stack of postcards "now." When you look at a galaxy, you are doing archaeology with photons. When you listen to thunder, you are measuring the distance to a lightning bolt that already happened. When you watch the Sun sink below the horizon, you are watching ancient light skim through kilometers of atmosphere and finally surrender to Earth's curvature. The universe is not synchronized to your awareness. It never was. What you call reality is whatever reality manages to deliver to you through a chain of physical processes that are slow, lossy, and noisy.

Now zoom in from galaxies to your daily life, and nothing changes except the scale. A friend speaks. Sound waves take time to reach you. Your ear turns pressure waves into neural pulses. Those pulses are filtered, compressed, compared to patterns, matched to meaning. Even your own body is a delayed report to yourself: pain signals arrive after damage; hunger arrives after chemical depletion; fear can arrive before conscious recognition because your brain runs a fast threat detector beneath your self-narration. The "present moment" is already a reconstruction by the time it feels immediate. You can prove this with a crude trick: look at a clock with a second hand while you blink. You will often feel as if the first tick after your blink lasted slightly longer, because your brain stitches continuity together. It post-processes. It back-fills. It smooths. It edits. The result is an internal broadcast that feels seamless, but it is not raw. It is the output of a pipeline.

That pipeline is the real protagonist of modern life. Not you, not your opinions, not your heroic self-image—your pipeline. Because access is not a philosophical privilege; it is a physical product. It is manufactured by sensors, channels, storage, and decoding. Every stage imposes limits. Every stage discards information. Every stage introduces delays. Every stage can be biased. That is why two people can stand in the same event and inhabit different practical realities: their pipelines are different. One has better sensors, better models, better attention, better timing. Another receives the key signal late, blurred, or never. The world is the same; the accessible record is not.

Think about how a giant scientific instrument actually "sees." A gravitational-wave observatory does not watch black holes collide the way your eyes watch fireworks. It measures tiny changes in distance—fractions of a proton's width—inside a long interferometer. The raw signal is buried under vibrations from trucks, microseisms from ocean waves, thermal jitter in materials, quantum noise in the lasers themselves. What the observatory produces is not a direct picture but a processed inference: after calibration, filtering, subtraction of known contaminants, cross-checking between distant detectors, and pattern matching against templates, a clean chirp emerges. Then, and only then, does the human story get told: two massive objects merged, spacetime rang like a bell, energy poured out as ripples. The collision did not become real when the chirp was announced. The collision was real when it happened. What changed was that a pipeline finally manufactured an accessible trace strong enough to survive all the noise.

Your life is full of smaller versions of that same structure. Your phone "knows" your location by combining satellite timing signals, antenna triangulation, inertial sensors, map constraints, and statistical smoothing. It is not measuring a single pure quantity. It is running an inference engine over partial, noisy cues. Your health tracker estimates sleep stages from a few signals and a model trained on population averages. Your car's driver-assist estimates lane boundaries from pixels, while glare and rain try to erase them. The trend is unmistakable: as the world gets more complex, direct perception gets replaced by pipelines that turn messy reality into legible summaries. And those summaries arrive with delay, with uncertainty, and with hidden assumptions. This is why access is never neutral. It is an artifact you manufacture, and the manufacturing process shapes what you believe is happening.

Three physical limits dominate every pipeline, from cosmic observatories to human senses. The first is bandwidth: how much information per unit time can pass through the channel. Your eyes have staggering bandwidth compared to a single text message, but even your eyes are bottlenecked—by the optic nerve, by attention, by the fact that your brain compresses and prioritizes. The second is noise: unwanted variation that contaminates the signal. Noise is not just random static; it includes everything you didn't measure, everything you can't control, everything that overlaps the same channel. The third is latency: delay between an event and the arrival of its trace in your accessible record. Latency is the silent assassin of certainty. A delayed truth is often treated as a new truth, because the mind mistakes arrival time for existence time. That error feels natural, but it is lethal in any environment where timing matters.

These limits force a brutal conclusion: the world you experience is not the world as it is. It is the world as your pipeline can deliver it. A pipeline is a sequence of transformations. Something happens in reality. A sensor transduces it into a signal. That signal is sampled—captured at discrete moments, not continuously. The samples are quantized—rounded into finite precision. The samples are compressed—details discarded to fit storage or bandwidth. The compressed signal is transmitted—delayed and sometimes corrupted. It is decoded—reconstructed imperfectly. It is interpreted—mapped onto a model of what could be happening. At every stage, information is lost. At every stage, bias can enter. At every stage, time passes. So by the time the signal becomes a belief in your head, it is already a filtered shadow of what occurred.

The most shocking part is that the pipeline doesn't merely report the world; it creates your sense of flow. Why does time feel like it has a direction? Because your accessible record grows in one direction. Memory is a physical storage process. It accumulates. It edits, but it accumulates. A creature that stores traces experiences a difference between "before the trace arrived" and "after the trace arrived." That difference becomes the felt arrow. The universe does not need a metaphysical arrow for you to feel one. Your record provides it. That is why the present feels special: it is the boundary between what your pipeline has delivered and what it has not yet delivered. The boundary is not written into spacetime like a cosmic fence. It is written into you as a limitation on access.

Once you accept this, a series of everyday mysteries become solvable. Why do people argue about what "really happened" after watching the same event? Because they didn't watch the same event; they watched different compressed, delayed, biased outputs. One person saw the beginning and missed the end. Another saw the end and inferred the beginning. One person's attention latched onto one cue and discarded another. Why does a late lab result cause a sudden change in behavior? Because the pipeline produced a data artifact with higher social weight—something that could be defended, recorded, acted upon—so belief shifted sharply. Why do warning dashboards often feel prophetic while also feeling useless? Because the pipeline for knowledge can remain strong even when the pipeline for influence is weak. Knowing is fed by incoming traces. Acting requires outgoing coupling. They are not the same channel. They do not share the same latency, the same noise floor, or the same bottlenecks.

This is also why "objectivity" is harder than it sounds. To be objective is not to float above the world. It is to engineer a pipeline whose distortions you understand, whose delays you can quantify, whose noise you can calibrate, whose blind spots you can audit. Science is not a magic mirror; it is a disciplined pipeline-building craft. You build instruments, you model their biases, you subtract contaminants, you test against null controls, you replicate under different conditions, you ask what would fool you. That is why the cosmic chirp is believable: because the pipeline's failure modes were hunted down and cornered. Everyday life rarely has that discipline. Most people run pipelines they did not design, on devices they cannot inspect, with models they do not understand, on channels that can throttle, reorder, and distort. That is not a conspiracy; it is just how complexity looks when it scales.

If the universe is an atlas of admissible histories, then your pipeline is the thin thread of access that crawls across the atlas, sampling it through a keyhole. The atlas can be structured, coherent, globally consistent, and still feel open and foggy, because you do not possess it. You possess a record. And the record is a product: it is manufactured by physical channels with limits. The deeper you go into modern life, the more you live through pipelines, and the less you should trust the feeling that what just arrived is what just became real.

In the next segment, the focus tightens on what happens when pipelines disagree—not as a social drama, but as a physical fact—and why "different access" creates different practical worlds even inside the same constraints.

Two people can stand under the same sky and still live in different universes. Not different in the poetic sense, but different in the literal operational sense: their accessible evidence differs, so the set of plausible stories they carry differs, so the future they are preparing for differs. We tell ourselves reality is a shared stage and disagreement is a psychological defect. But a deeper truth is simpler and colder: disagreement is often just mismatched access. The world is one; the pipelines are many. And because the pipelines are many, "the present" splinters into multiple private versions of now.

Astronomy is the cleanest place to see this because it strips away ego. Put two telescopes on the same mountain. Point them at the same galaxy. One telescope is large, cooled, calibrated, equipped with adaptive optics and a spectrograph. The other is small, warm, noisy, with a cheap detector. The first sees structure: spiral arms, dust lanes, star-forming regions, redshifts that reveal motion, chemical fingerprints that reveal composition. The second sees a blur. They are "looking at the same object," yet one pipeline delivers a richly structured record and the other delivers a smear. The galaxy did not become more detailed for one observer. Access did. The first observer's universe contains a rotating disk with measured dynamics; the second observer's universe contains a fuzzy patch. The same reality, two different accessible worlds.

Now scale that down to human life, where the pipelines aren't just telescopes but entire networks of sensors, filters, and models. You and your neighbor both live through weather, but your access differs. One of you has a high-resolution forecast, a barometer, an understanding of pressure fields, a habit of checking radar. The other has a glance out the window and a vague feeling in the knees. When the storm arrives, one says it was obvious, the other says it came out of nowhere. Neither is lying. Their records were different. Their confidence was different. Their practical futures were different. And if you've ever watched a serious weather event unfold, you've seen the strange social fact that follows: the world becomes "real" for the public only when the rain is already on their face, even though the radar was already screaming. That is not stupidity. That is access distribution.

The same structure appears in health, and it becomes emotionally violent because the stakes are personal. A body can drift toward crisis while the person feels "fine," because the earliest signals are invisible to casual perception. Meanwhile a clinician with instruments sees faint but meaningful deviations—oxygen saturation, subtle patterns in labs, a rising marker, a change in variability. Again: the body didn't change when the number appeared. The record did. But now add the truly modern twist: the record itself can disagree across pipelines. A wearable says you slept well; you feel exhausted. A pulse oximeter reads low because your finger is cold; another reads normal. A measurement is delayed; the trend is hidden; the picture is patched together from gaps. Two observers can look at the same person and inhabit different possible corridors because each has a different slice of evidence. That is why fights break out over what is "really happening." Not because reality is subjective, but because access is fragmented.

Fragmentation becomes more dramatic when the pipelines are not passive, but actively shape what you see. Consider a feed of information—news, messages, alerts, images—delivered through a filtering process. The filtering can be as simple as your own attention or as complex as recommendation machinery that decides what you see based on what keeps you engaged. In such a world, "public reality" is not a single broadcast. It is a swarm of tailored streams. Two people can live in the same city, during the same week, and still receive evidence that pushes them toward different narratives about what is happening. Then they meet and cannot understand how the other could possibly believe what they believe. They treat it as madness or malice. But beneath the insult is a physics-like truth: their pipelines sampled different slices of the world. Their records are not the same. Their priors were fed different inputs. The disagreement is not just ideological; it is informational.

Even your own mind contains pipeline disagreements. Your fast threat circuitry can react before your conscious story catches up. Your gut can tighten before you can name the reason. Your attention can lock onto one cue and ignore another. Your memory can revise the order of events to make the story cleaner. You experience your mind as one coherent narrator, but that narrator is the last stage of a layered processing stack. Some parts of you live in a different "now" than other parts of you, because some signals are fast and crude and others are slow and detailed. That is why you can feel safe while being in danger, or feel danger while being safe. Access is not only external; it is internal. The pipeline is not a single tube; it is a braid.

This leads to a ruthless conclusion: the shared world is not what people argue about; it is what their records allow them to argue about. If you want to understand conflict, don't start with psychology. Start with what evidence each person actually received, when they received it, how noisy it was, and what filters shaped it. Timing matters because access arrives in sequence. A person who receives a warning early carries a narrowed corridor early. A person who receives the same warning late carries a broad corridor longer. The first will appear alarmist to the second, and the second will appear careless to the first, even if both are rational given their record. This is one of the hidden engines behind the feeling that modern life is full of "unforced errors": the warnings exist, but they are unevenly distributed in time and credibility.

Credibility is itself a channel property. Some evidence arrives in a form that can be defended: a number, a timestamp, a signature, a verified measurement. Other evidence arrives as intuition, rumor, vague unease. The world doesn't care about defensibility, but humans do. They are social creatures who must justify actions to other humans. So the practical weight of evidence is not just about its truth; it is about its format. This creates a bizarre distortion: an early signal can be accurate and still ignored because it arrives in a weak social format, while a later signal can be less informative and still drive action because it arrives with authority. A graph triggers response; a gut feeling triggers ridicule. Again, this is not a moral story. It is the physics of how evidence becomes actionable in groups.

Now add another layer: the pipeline does not only deliver facts, it delivers uncertainty. A good instrument does not merely output a value; it outputs a value with error bars, calibration notes, and context. A poor instrument outputs a clean number with hidden bias. The second feels comforting and can be dangerously persuasive. This is why sophisticated access can look weaker than crude access. A careful scientific pipeline often says, "Here is what we think, here is how confident we are, here is what could fool us." A sloppy pipeline shouts, "This is what happened," with no qualifiers. People often prefer the shout because it feels like reality. But reality is not obligated to be clean. The clean shout is often a lie manufactured by a pipeline that discards uncertainty to seem decisive.

In a constrained universe, uncertainty is not an embarrassment; it is a signature of partial access. You can often tell how honest a pipeline is by how it handles uncertainty. Does it admit noise? Does it admit delay? Does it admit missing data? Or does it pretend the world is perfectly readable? The Volumetric Time Model is not impressed by perfect readability, because perfect readability is rarely physical for embedded agents. The model expects partial access. It expects blind spots. It expects disagreement. And it treats the growth of the record—more measurements, more evidence, more constraints—as the mechanism by which belief corridors narrow over time.

This also explains why people sometimes feel reality "flips" abruptly. In truth, the underlying process often changes smoothly, but access crosses a threshold. The evidence arrives in a form that collapses the plausible corridor set. Before the threshold, multiple stories fit. After the threshold, one story dominates. The subjective experience is a sudden switch: yesterday uncertain, today obvious. But the switch is in the observer, not in the world. The world was already moving along a corridor shaped by constraints.

The observer's pipeline simply didn't deliver the decisive clue until now.

Once you see this, you can predict a peculiar modern pathology: a society can become extremely good at generating forecasts while becoming bad at coordinated timely response, because forecasting is mostly about reading incoming traces, while response is about coupling action outward through delayed, noisy pathways. The incoming channel can improve with sensors and models. The outgoing channel can remain bottlenecked by friction, delay, and competing influences. The result is a civilization that can see dangers coming and still fail to steer away. That failure then produces moral panic, accusations, and myths—because it feels intolerable to admit a simple truth: seeing is not the same as turning.

So when two people fight about what is "really happening," it is worth remembering that they are not necessarily fighting over the world. They are fighting over their pipelines. They are fighting over which evidence is allowed into the record, which evidence is dismissed as noise, which formats count as credible, and which delays are tolerated. In that sense, the deepest struggle in modern life is not over reality but over access to reality. Not because reality is subjective, but because access is physical, uneven, and engineered.

A pipeline is a machine that turns the universe into something you can hold. That sounds poetic until you remember what a pipeline really does: it throws most of the world away. Not because it's evil, but because it has to. The raw world is too large, too fast, too continuous, too hungry. If you tried to ingest reality at full resolution, you would drown in photons and noise. So every observer—human, animal, instrument—does the same three brutal moves: it samples, it compresses, and it interprets. Sample: take a continuous world and chop it into discrete readings. Compress: discard detail to fit bandwidth and storage. Interpret: map the leftover scraps onto a model so it becomes meaning instead of static. Your "reality" is the output of those three moves, running nonstop, under constraints you did not choose.

Sampling is the first cut, and it is where the world starts to fracture into artifacts. A camera does not "see" motion; it grabs frames. An audio recorder does not "hear" continuity; it grabs snapshots of pressure and rebuilds a waveform. A medical monitor does not watch a heart as a living thing; it samples electrical potentials, filters them, then displays a pattern you can recognize. Sampling creates a hidden danger: aliasing. If you sample too slowly, high-frequency truth masquerades as low-frequency lie. A wheel spins and appears to rotate backward. A flickering light appears steady. A fast instability looks like a slow drift. Nature didn't deceive you; your pipeline undersampled the signal. In ordinary life, aliasing looks like misjudgment: you check something once a day and conclude it's stable, not realizing it's oscillating wildly in between. You look at a relationship once a month and conclude it's fine, not realizing the cracks are propagating daily. You sample the world too sparsely and mistake the gaps for calm.

Then comes compression. Compression is not merely shrinking data; it is choosing what to forget. Your eyes compress brutally. You have a high-resolution fovea and a blurry periphery because you cannot afford full resolution everywhere. Your brain compresses even more brutally: it builds objects, categories, narratives, and then stores the narrative rather than the raw sensory flood. You don't remember every leaf; you remember "a tree." You don't remember every syllable; you remember "an insult." Compression is why you can function, but it is also why your inner broadcast feels clean. Clean broadcasts are usually heavily compressed broadcasts. The mess was thrown away upstream.

Instruments compress too, and the modern world is built on that fact. A telescope doesn't hand you every photon; it bins counts, subtracts background, corrects for instrument response, and delivers a picture that fits on your screen. A radio dish doesn't hand you the raw electromagnetic field; it integrates, filters, and reconstructs. A brain scanner doesn't hand you a mind; it hands you a proxy constructed from signals that are already indirect. Every famous "image" in science is really a reconstruction—an inference product dressed in visual clothing so your human cortex can understand it.

The third move—interpretation—is where the pipeline turns from a passive channel into an active sculptor. Interpretation means you don't just store data; you translate it into a belief about what is happening. You match patterns. You fit curves. You infer hidden variables. You decide what counts as signal and what counts as noise. And here's the part people don't like to admit: interpretation always uses a prior story. Even when you think you are "just looking," you are looking through expectations about what kinds of things exist and what kinds of changes are plausible. If you expect a planet, you search for periodic dimming. If you expect a tumor, you search for a suspicious contrast pattern. If you expect betrayal, you search for a certain tone in a voice. Priors are not weakness; they are survival. Without priors, the world is an ocean of unranked possibilities and you cannot act.

This is why two pipelines can stare at the same raw input and produce different practical realities. One pipeline has a prior that says, "This kind of fluctuation is normal drift." Another has a prior that says, "This kind of fluctuation is the start of runaway instability." One pipeline treats a pattern as noise; another treats it as a signature. That difference can come from training, from experience, from better tools, or from bias. In the laboratory, we try to discipline priors with calibration and replication. In daily life, priors are often fed by habit, fear, and social contagion. The result is that the world's raw signals are the same, but the reconstructed world in each mind is different.

Now, the most dramatic pipeline lesson in all of physics is gravitational-wave detection, because it shows you what "reality" looks like when it is almost entirely buried. The signal is a whisper in a hurricane. The detectors are shaken by the Earth, by thermal agitation, by quantum jitter in the laser light. You don't "see" two black holes collide. You infer it by building a pipeline that knows what the collision's fingerprint should look like, then searching for that fingerprint across noisy data, then cross-checking between separate detectors, then ruling out mundane sources, then demanding statistical confidence that the pattern isn't accidental. The public receives the final chirp and thinks of it as a message from the cosmos. But the chirp is a manufactured artifact: a refined trace produced by an engineered pipeline that throws away enormous quantities of irrelevant motion to reveal a faint coherent pattern.

That is how your daily access works, too, except you don't see the error bars. You don't see the discarded frames. You don't see the assumptions baked into the reconstruction. Your mind hands you the result as a smooth movie called experience. The movie feels immediate and authoritative because it has to. An organism that constantly doubts its own sensory feed doesn't last long. So confidence is often a user interface trick: the pipeline outputs a story with high subjective certainty even when the underlying inference is fragile.

This matters because fragility is where modern life hides. Many of the most important variables are not directly observed. They are inferred. And inferred variables can stay "stable" in your story until they cross a threshold, at which point the pipeline snaps into a new interpretation and you experience it as a sudden change in reality. But the underlying process often drifted for a long time. The pipeline simply lacked the resolution, or the right features, or the right trigger, to revise its story earlier. This is why so many people describe disasters as "out of nowhere." Out of nowhere is often just out of your pipeline.

There is another pipeline trick that deserves a spotlight: smoothing. Smoothing is the art of averaging away volatility so you can see a trend. It is essential. It is also dangerous. Smooth too aggressively and you erase early warning spikes. Smooth too little and you drown in noise. The choice of smoothing is a choice about how you trade false alarms against missed alarms. It is a choice about how you experience time. A heavily smoothed feed makes the world feel calm until it suddenly isn't. A lightly smoothed feed makes the world feel chaotic even when it is stable. Your emotional relationship with reality can be altered by a single dial in the pipeline.

And pipelines don't just smooth the world; they discretize it into categories. Categories are compression. Categories are also weapons. Once a continuous spectrum is binned into labels, subtle changes disappear until a label flips. You go from "normal" to "abnormal" as if a cliff exists, when the underlying variable changed gradually. You go from "safe" to "unsafe" as if an invisible boundary was crossed, when the boundary is often an artifact of measurement resolution and thresholds. Nature is continuous; your access is quantized.

In a constrained atlas view, all of this becomes less mystical and more mechanical. The world within a bounded region follows rules and constraints. Those rules carve structure. Structure makes it possible for partial evidence to be informative. But what you actually receive is a compressed, delayed, noisy record shaped by a pipeline. Your uncertainty is not the universe hesitating; it is your pipeline failing to isolate a single corridor early. Your certainty is not a metaphysical revelation; it is a collapse of plausible corridors after a decisive trace finally arrives in usable form. Your sense of "flow" is not a cosmic river; it is your record accumulating one packet at a time.

Once you see the observer as a pipeline, you stop blaming yourself for not being omniscient and you start asking sharper questions. What is my sampling rate? What am I missing between samples? What is my noise floor? What assumptions are baked into my reconstructions? Which facts arrive quickly and which arrive late? Which signals arrive in a form that carries authority and which arrive as vague unease? Which channels are throttled, distorted, or biased? These questions don't require metaphysics. They require engineering-level honesty about how reality becomes legible to an embedded creature.

The unsettling punchline is that improving access is not the same as becoming wise. You can build pipelines that deliver more data and still mislead you if the interpretation layer is wrong. You can build pipelines that deliver crisp stories with high confidence and still be wrong because they suppress uncertainty. You can build pipelines that see far ahead and still fail to help you act because access and influence travel through different channels with different timing. The universe is not obligated to align those channels for your convenience.

What you call reality is the intersection of two things: what exists within constraints, and what your pipeline can actually deliver to you before it's too late to matter.

A pipeline can fail in two ways that look identical to a human: it can miss a real signal, or it can hallucinate one. Both failures feel like "reality," because you never see raw reality. You see the output. This is why the most dangerous thing a pipeline can do is not to be wrong—it's to be confidently wrong. A pipeline that admits uncertainty is annoying. A pipeline that prints a clean story with a clean number can change lives, move crowds, shift markets, start wars, end careers, trigger panic, cause paralysis. And the tragedy is that the confidence often comes from the very same trick that makes pipelines useful: compression. When you compress, you throw away ambiguity. You replace the messy cloud of possibilities with a crisp label. The label feels like truth, even when it's just a convenient summary of incomplete evidence.

In the history of science, entire eras were shaped by this problem. Astronomers once thought they saw canals on Mars. Not because Mars grew canals, but because a pipeline—telescopes with limited resolution, eyes with pattern hunger, minds trained to see lines—turned blurry shading into engineered geometry. The canals became a story that fit the compressed output. Then better optics arrived, better calibration, better imaging, and the story evaporated. Mars didn't change. Access changed. The pipeline stopped producing the illusion. The lesson is brutal and permanent: if you want to trust your view of reality, you must understand how the view is manufactured.

Modern astrophysics is filled with elegant examples of pipelines fighting their own hallucinations. Consider how we find planets around other stars. We don't usually see the planet directly. We see a tiny dimming of starlight, repeating with a regular rhythm, like a moth passing in front of a floodlight. But dimming can come from many things: starspots rotating, binary companions, instrumental drift, cosmic rays striking a detector pixel, processing artifacts introduced by filtering. The raw signal is fragile. The pipeline must be trained to distrust itself. It must ask: if I scramble the time stamps, do I still see the pattern? If I look at a different star, do I see the same fingerprint? If I split the data into segments, does the rhythm persist? If I change the extraction method, does the candidate survive? A planet is not proclaimed because a curve looks pretty; it is proclaimed because an entire battery of self-skeptical tests fails to kill the candidate. That is what makes it real in a scientific sense: not romance, but survival against null attacks.

The same logic appears in particle physics. A detector sees flashes, tracks, deposits of energy. But the laboratory is flooded with background: random coincidences, electronic noise, stray radiation, miscalibrations, drifting thresholds. The pipeline is a courtroom with a hostile judge. Every claimed discovery is cross-examined by null controls: run the experiment in a configuration where the signal should vanish; randomize labels; shift timing; measure background in sidebands; look for the same signature where it should not exist. The famous discoveries survive because the pipeline cannot easily manufacture them under conditions designed to make hallucinations likely.

Gravitational-wave detection pushes this to near absurdity. The sought signal is so faint that almost any environmental vibration can mimic it. So the pipeline becomes an obsessive skeptic. It demands coincidence between separated detectors. It demands consistency in waveform shape. It demands that the signal does not appear when data is time-shifted in unphysical ways. It demands that known instrumental artifacts be ruled out. In other words, it doesn't trust the chirp because it's beautiful. It trusts it because every attempt to create the chirp in a null world fails.

A null world is the key concept here, and it is not limited to laboratories. A null world is a version of your data where the relationship you think you see should not exist. If the relationship still appears, your pipeline is manufacturing patterns. If it disappears, you have evidence that the relationship might be real. This is the foundation of disciplined inference in any complex environment: you must build null worlds that preserve the boring structure of your data—autocorrelation, seasonal cycles, measurement noise—while destroying the specific connection you're claiming. Otherwise you are not testing your claim; you are testing your ability to fool yourself.

Human life, unfortunately, is filled with pipelines that never build null worlds. They take a pattern, feel a thrill, and call it meaning. A person checks a dashboard that seems to predict a crisis, and they believe the dashboard is prophetic without asking whether the "prediction" is just leakage—an early trace of the crisis already present in the inputs. A trader sees a signal and believes it is causally powerful without asking whether it is merely correlated with an underlying driver. A medical model flags risk and gets credit for prediction without asking whether the flag is simply reading the earliest symptoms already unfolding. These are not trivial distinctions. They are the difference between knowing and steering. A pipeline can be excellent at forecasting while being useless for intervention. It can be a perfect mirror and still be a terrible lever.

This is where the observer-as-pipeline idea becomes more than a description; it becomes a discipline. If you want to talk about access honestly, you must talk about failure modes. There are a few classics that return in every field like stubborn ghosts.

One is confounding: a hidden variable drives both what you observe and what you think you're influencing. You act, something happens later, and you credit your action, but in truth both your action and the outcome were downstream of the same hidden driver. Your pipeline mistakes shared cause for control. This is the oldest illusion in human history: we perform a ritual, then the rain comes, therefore the ritual caused the rain. In a modern setting, the ritual might be an intervention policy, a trade, a recommendation tweak, a protocol change. The illusion persists because the pipeline lacks a clean way to isolate causal coupling from shared structure.

Another is leakage: information from the future sneaks into the present through your measurement process, not through physics. A model "predicts" an event because the data it is allowed to see already contains post-event traces—labels inferred from outcomes, summaries computed after the fact, signals that are only recorded because the event happened. Leakage is an accounting error, but it can look like magic. The cure is to enforce causality in the data pipeline: only allow features that truly exist at the time of prediction. In real life, this is harder than it sounds, because logs arrive late, timestamps are messy, and human behavior changes what gets recorded.

Another is non-stationarity: the world changes while your pipeline assumes it doesn't. A calibration that held last month fails this month because the environment shifted. A sensor drifts. A population changes. Incentives mutate. Your pipeline continues to interpret today's data with yesterday's mapping, and it doesn't notice until it collapses. This is why pipelines need continuous recalibration, and why any claim about reliable access must include a story about drift and adaptation.

Another is threshold illusion: continuous reality is binned into discrete labels, so nothing looks wrong until the label flips. A person is "fine" until "critical." A market is "stable" until "crash." A relationship is "normal" until "over." The underlying variable can drift steadily while your pipeline prints the same category. Then, suddenly, it prints a new category and you experience it as a jump in reality.

The jump is a quantization artifact. The world moved continuously; your access was discretized.

These failure modes matter because VTM is trying to say something precise about what an embedded observer can know, and later what an embedded observer can do. If your access is delivered by pipelines that can hallucinate, then your belief over possible histories can collapse onto the wrong corridor with terrifying confidence. You will feel certainty while being wrong. You will act with conviction while being misaligned to reality. That isn't metaphysics. It is what happens when interpretation outruns calibration.

So a serious account of access must include a serious account of how to test access. The most basic testing move is the null attack: destroy the claimed relationship while preserving everything else, and see if the pipeline still produces it. If it does, you are looking at an artifact. If it doesn't, you may be seeing a real coupling. In time-ordered data, null attacks can be done by shuffling within appropriate blocks, or by shifting sequences in a way that preserves internal correlation while breaking cross-coupling. Another essential move is stratification: if a signal appears only because you mixed different regimes together, then you didn't find a law; you found a mixture artifact. You must separate regimes, test within them, and see what survives.

This is the deeper reason to treat the observer as a pipeline rather than a magical witness. A magical witness is always right by definition. A pipeline can be audited. A pipeline can be stress-tested. A pipeline can be attacked with null worlds designed to reveal hallucinations. Once you start thinking this way, you stop asking childish questions like "Is the future real?" and you start asking adult questions like "What traces of future corridors leak into present observables under these constraints, and can my pipeline detect those traces without being fooled?" You stop worshipping predictions and start interrogating their provenance.

And there is a final twist that makes this worldview feel almost like a new kind of realism: the better your pipeline becomes, the more it should talk about its own limits. True access comes with error bars. True access admits blind spots. True access knows what could fool it. A pipeline that never admits uncertainty is not strong; it is theatrical. In the real universe, embedded observers do not get perfect feeds. They get partial traces. The honest observer is the one who can say not only what they think is happening, but why they think it, what would falsify it, and how close they are to the noise floor.

Once you build that discipline, you can finally approach the most unsettling regime with clarity: the regime where the pipeline can forecast with real statistical strength, while the channel for influence has already faded below detectability. A world can be legible and still untouchable. That is not despair; it is a measurable property of delayed coupling. And it is where the story stops being philosophical and starts being experimental.

There is a quiet terror in discovering that your pipeline can be brilliant at seeing patterns while being useless at changing outcomes. Humans carry a comforting superstition: if you know enough, you can do enough.

Knowledge feels like power because, on small scales with tight feedback, it often is. You touch a hot stove, pain arrives quickly, you pull your hand away, and the world responds instantly. That childhood loop trains the brain to fuse seeing and steering into one sensation: awareness equals control. But the universe is under no obligation to preserve that fusion. Once delays stretch, once noise swells, once the relevant levers become indirect, knowledge and influence separate like two diverging beams of light.

The first beam is incoming: traces of what is happening in the world arrive to you. That beam can become astonishingly sharp. A modern observer can carry a pocket rectangle that receives satellite imagery, global measurements, patterns detected by machines, warnings extracted from rivers of data. Even without any machine, a trained mind can infer hidden structure from subtle cues because constraints carve ridges into reality. Incoming access can be upgraded by better sensors, better models, better attention, better calibration. That is why forecasting can keep improving. The universe leaks the future into the present, not by magic, but by physics: inertia, coupling, conservation, feedback, and constraints. The present cannot help but contain fingerprints of what it is becoming.

The second beam is outgoing: your actions attempt to couple back into the world and leave a signature that persists long enough to alter what happens later. That beam is easily degraded. It is degraded by delay (your push arrives late), by noise (other influences swamp your push), by indirectness (you can't touch the relevant variable directly), by bandwidth limits (you can't act frequently or precisely enough), and by energy limits (coherent influence costs resources). This is why powerlessness can be a physical condition even in a world that is highly legible. You can have a telescope that sees a storm forming at sea while having no ship, no shelter, no lever to move the air. You can have a dashboard that predicts collapse while having no way to inject timely influence into the fragile variable that would avert it.

To feel this separation, imagine standing on a cliff watching a glacier calve. You can see the cracks long before the ice breaks. You can hear the groaning. You can measure vibration, temperature, strain. Your incoming access can be exquisite. But your outgoing influence is close to zero. You cannot push the glacier into stability with your hands. Even if you had machinery, you would still be fighting inertia and scale. The world is legible; it is not steerable by you. There is nothing mystical about this. It is a ratio of influence to background and a mismatch of timescales.

The modern twist is that we live surrounded by prediction artifacts that make the first beam feel omnipotent. A model produces a probability. A curve points upward. A warning flashes. The human mind interprets this as a kind of ownership: if I can foresee it, I should be able to prevent it. But foreseeing is often just reading a trace that is already present. The future corridor is already tightening, and the model is simply sensitive enough to detect the tightening early. That sensitivity is real. It is also not the same thing as leverage. A thermometer can forecast fever. The thermometer cannot treat the fever. A seismograph can detect tremors. The seismograph cannot hold the tectonic plates still. A high-gain microphone can hear the early whisper of instability. The microphone cannot silence it.

This is why a disciplined observer must separate two questions that people constantly fuse: "Can I infer what is likely to happen?" and "Can I make a detectable difference to what happens?" The first is about extracting structure from the incoming beam. The second is about surviving the decay of the outgoing beam. If you fail to separate them, you will spend your life mistaking legibility for control, and every time the world ignores your effort you will interpret it as personal failure or cosmic cruelty. But often it is neither. It is simply that your influence falls below the threshold at which the world can register it as anything other than background.

Thresholds matter because the world is not a courtroom. It does not award partial credit for intention. It doesn't care that you tried. If your action does not create a difference that survives into later observables, then for all practical purposes it did not steer the outcome. This is painful to accept because humans want moral continuity: effort should matter. But physics is indifferent to effort. Physics cares about coupling. A whisper can steer a sensitive resonance if timed perfectly; a shout can fail if injected too late. Timing is not romance here; it is geometry. A push placed within the window of sensitivity can redirect a trajectory. The same push outside that window disappears into noise.

This is why delays are so lethal. Delay turns action into an echo. You act, but the world receives your action later, after it has already moved. The longer the delay, the more the world changes in the meantime, and the more your action becomes mis-timed. Mis-timing reduces coupling. Reduced coupling makes your influence look like noise. Noise is indistinguishable from "nothing happened." This is not a metaphor; it is what every engineer learns when they try to control a process with delayed feedback. You can have perfect knowledge of the process and still fail to control it if the loop closes too late.

Noise completes the trap. Noise is everything else happening while you act. In a clean environment with low noise, small actions produce visible differences. In a high-noise environment, your action must be large or exquisitely timed to rise above the background. If the background is massive—many competing influences, hidden variables, chaotic amplification—your action can be swallowed. This is why influence is not evenly distributed in the universe. Some domains are low-noise and tightly coupled, so small agents can steer. Other domains are high-noise and loosely coupled, so only large interventions matter. Many of the most important modern processes live in the latter category: complex, delayed, saturated with competing forces. They are readable because data is abundant. They are hard to steer because coupling is diluted.

This separation also explains a subtle cruelty in the human experience of warning. A warning is a form of access. It sharpens your belief corridor. It can arrive early enough to make you feel responsible, yet late enough that your remaining lever is weak. Then you are trapped in a psychological double-bind: you can see the cliff well enough to blame yourself for the crash, but you cannot generate a steering signal strong enough to avert it. This produces the modern sensation of watching disaster approach in high definition. The tragedy is not ignorance; it is legible powerlessness.

And legible powerlessness breeds myths. When people cannot steer, they often adopt stories that dissolve the pain: fate, inevitability, "it was meant to happen." Those stories are emotional anesthesia. They may even be socially useful. But the more interesting explanation is physical: the outgoing beam decayed below threshold before the incoming beam lost clarity. Knowledge stayed strong; influence died. In that regime, the mind is forced to witness without agency, and witnessing without agency is unbearable for creatures trained on tight feedback.

Now return to the atlas picture. Inside a bounded region, constraints define admissible histories. Your record delivers partial evidence about which corridor you occupy. As evidence accumulates, your belief narrows. That's the incoming beam tightening your corridor. But your reachable corridor set—the subset you can actually steer into—can shrink at the same time. It shrinks because the levers you can pull are delayed, indirect, noisy, limited in bandwidth. So you can arrive at a future with high confidence and low influence: you know where the corridor leads, but you cannot shift into a neighboring corridor because the door is already closing. The closer you get to the boundary, the more your remaining influence resembles a whisper in a hurricane.

This is where the observer-as-pipeline idea becomes deeply practical. A pipeline is not just a way of seeing; it is also a way of acting. The outgoing beam is also a pipeline: decision, execution, propagation, effect, measurement. Every stage adds delay and noise. Every stage can be throttled. Every stage can be distorted. When people talk about "willpower" as if it is a single internal knob, they ignore the fact that action is a chain. If any link in that chain is delayed or degraded enough, the entire chain fails to couple. The failure is then misinterpreted as weakness. But the pipeline view reveals the colder reality: your action channel may be physically mismatched to the process you're trying to steer.

A final cosmic image clarifies why the separation between seeing and steering is not an exception but the default. Light travels fast, but even light has a finite speed. Gravity travels at the speed of light. Signals take time. Influence propagates. You can observe a distant supernova long after it exploded. Your observation does nothing to stop the explosion. That feels obvious because the event is far away. Yet the same structure exists in your daily life whenever the relevant coupling is slow or indirect: you receive evidence of an outcome long before you can alter it, or you can alter it only weakly compared to the forces already in motion. The universe is full of such mismatches. The myth is that humans are exempt.

When you accept that, you stop treating the separation as existential horror and start treating it as measurable architecture. There are regimes where access improves faster than influence decays, and those are the regimes where foresight becomes power. There are regimes where influence decays faster than access improves, and those are the regimes where foresight becomes torment. The crucial point is that both regimes are physical. They depend on delay, noise, bandwidth, scale, and constraint structure. And once you see that, you can ask the only question that matters in a world of pipelines: not "Is the future real?" but "When does influence drop below detectability, even while prediction remains above it?" That boundary is where modern life truly begins.

Now comes the question that separates a bedtime philosophy from a usable theory: how do you tell whether you actually have access, and how do you tell whether your actions actually couple? Because the universe is full of situations where your pipeline gives you a gorgeous narrative while giving you no leverage, and it's also full of situations where you think you're steering when you're really just riding a current. If you want the observer-as-pipeline idea to be more than a mood, you need a discipline for interrogating the pipeline. Not just "What do I believe?" but "What was I allowed to observe, how delayed was it, how noisy was it, what could have fooled me, and what would count as proof that my action mattered?" This is where the story stops being mystical and becomes almost engineering-like: you treat access and influence as two different couplings and you test them the way you'd test any fragile connection in nature.

Start with access. Access is not "having data." Access is having data that is timely, relevant, and strong enough to narrow the set of plausible histories before the decision window closes. A pipeline can drown you in data while starving you of access. You can have a thousand numbers and still be blind if the numbers are delayed, if they measure proxies rather than drivers, if they are smoothed until they erase early warning, or if they arrive in formats too weak to trigger action. You can also have access with very little data if the data has the right structure—one clean trace that rules out most corridors. In physics, a single spectral line can tell you what a star is made of. In medicine, one marker can reveal an underlying cascade. In navigation, one landmark can tell you which road you're on. Access is about the power of evidence, not the volume of it.

So how do you test whether your pipeline is actually delivering access rather than theater? The first move is brutally simple: ask what would look the same if the story were false. This sounds obvious, but most human pipelines never do it. They see a pattern and stop. A disciplined pipeline asks, "Could this pattern arise from noise, from drift, from bias, from my own filtering choices?" In science, we answer that by building null worlds: versions of the data where the relationship you think you see should not exist, while everything else—sampling rate, noise texture, daily cycles, autocorrelation—remains similar. If your pipeline still "sees" the relationship in the null world, you are not seeing a law; you are seeing a hallucination produced by the pipeline itself.

Time-series data makes this especially treacherous because time-series data has memory. A signal today is often correlated with itself tomorrow. That internal correlation can fool you into thinking you've found a predictive relationship between two things, when you've really just found the persistence of one thing. A cheap pipeline shuffles data randomly and claims victory when the effect disappears, but random shuffling often destroys the internal memory structure, creating a null world that is too easy to beat. A serious pipeline builds null worlds that preserve the internal texture of each stream while breaking the cross-coupling between them. One way is to shift one stream in time relative to the other by an amount large enough to destroy true coupling while keeping each stream's internal rhythm intact. Another is to shuffle within blocks so the short-term structure survives but the alignment you care about is broken. The details matter, but the philosophy is clean: you test your pipeline against opponents that share the same boring structure, so you don't confuse "I can beat a toy null" with "I have access to reality."

The second move is to ask whether your access survives changes in the pipeline itself. If a tiny change in smoothing window, threshold, preprocessing, or feature extraction flips your conclusion, then your access is fragile. Fragile access is not worthless, but it is not the kind of access you should bet your life on. In astrophysics, if a planet candidate vanishes when you change the detrending method, you don't declare a planet; you declare the pipeline is sensitive. In everyday life, people do the opposite: they find the one preprocessing choice that makes the story look clean and then call the story truth. A disciplined observer does not trust the story that depends on one perfect knob setting. Reality is not obligated to align with your favorite knob.

Now shift to influence. Influence is not "I did something." Influence is "my action left a detectable signature later, above what would have happened anyway." That "above what would have happened anyway" clause is where nearly all human self-deception lives. You act, the world changes, and you take credit. But the world was going to change. Or worse: the same hidden driver that pushed you to act also pushed the outcome. You feel like a steering wheel when you were actually a passenger with good intuition about bumps in the road. Again, science has a cure: you build comparisons that isolate the action's signature. You ask: if I randomize the action timing, does the outcome coupling persist? If I hold the action constant, does the outcome still drift the same way? If I compare cases where the action differed but everything else was similar, does the distribution of outcomes shift? If you cannot build such comparisons, then claims of influence are usually stories, not measurements.

In real life, it's hard to run clean trials, but the conceptual point still matters: influence is a property of a channel from action to later observables. Channels degrade with delay and noise. The longer the delay, the more opportunities the world has to wash out your signature. The higher the noise, the more your signature is submerged. The more indirect the lever, the more your action disperses across variables that don't matter. This is why you can "do the right thing" and still see no effect: your action may be correct in principle but below threshold in practice. The universe does not reward moral correctness; it responds to coupling strength.

There is an especially cruel case that modern life produces constantly: your knowledge pipeline improves faster than your influence pipeline. Sensors get better. Models get better. Forecasts get sharper. Meanwhile, the pathway from decision to effect becomes longer, more delayed, and more crowded with competing influences. The result is a civilization that produces increasingly accurate early warnings while feeling increasingly impotent. People then become spiritually nauseous. They swing between arrogance ("we can predict everything") and fatalism ("nothing can be changed"). But the sober diagnosis is neither arrogance nor fatalism. It is asymmetry between two channels. Incoming traces remain readable. Outgoing influence becomes diluted.

This asymmetry can be tested, too, and the test has the same flavor as the access test: build null worlds for influence. If you suspect your action matters, ask whether the measurable coupling between your action and later outcomes remains when you destroy the causal alignment. If the coupling remains in the null world, then the observed dependence is likely shared structure, not steering. If the coupling collapses under null attacks while prediction dependence remains strong, you are staring at the most interesting regime of all: the world is legible, but your influence is below detectability. That regime is psychologically unbearable, which is why humans try to deny it, but it is physically plausible, and in complex delayed environments it is common.

At this point, the observer-as-pipeline idea starts to sound like a new kind of realism. It says: stop treating "truth" as a thing you possess and start treating it as a thing you approximate with finite channels. Stop treating "power" as a moral property and start treating it as the ability to maintain coupling through delay and noise. Stop treating "surprise" as reality changing and start treating it as your pipeline receiving a late decisive trace. Stop treating "certainty" as cosmic endorsement and start treating it as your belief corridor narrowing because your record improved. This is not a cynical worldview. It is a worldview that refuses to flatter the observer. It demands you face the fact that you are a bounded creature inside a bounded region, reading the world through a keyhole.

And once you accept that, you can build a practical habit that is almost like a personal scientific method: an access audit. Not paperwork, not ritual—an audit in the literal sense of checking your channels. What is my latency? What is my noise floor? What variables am I truly observing versus inferring? What preprocessing am I applying without admitting it? What would I still believe if this data stream were shifted, scrambled in blocks, or otherwise stripped of the alignment I'm relying on? If my conclusion vanishes under small pipeline changes, am I looking at a real feature or a fragile artifact? If I act, what observable signature would convince me my action mattered, and what null comparison would falsify that claim? These questions are not academic. They are survival questions in a world where your mind is constantly tempted to confuse a clean story with a clean reality.

The grand lesson is this: the universe is not obligated to be readable in the same places it is steerable. Your pipeline can deliver beautiful foresight and still leave you with no leverage. The only honest response is not despair, and not magical thinking, but measurement: test access against null worlds; test influence against null worlds; separate incoming traces from outgoing coupling; and learn to recognize the regime where knowledge remains above the noise while influence has already sunk beneath it. That regime feels like a curse, but it is also the key to understanding why modern life can feel simultaneously brilliant and powerless—like standing in a planetarium with perfect projections while the real sky outside is doing whatever it wants.

The final trap is the one your mind loves most: it confuses a good story with a good channel. Humans are narrative engines. We would rather have a coherent explanation than an accurate one, because coherence is psychologically calming. A pipeline that outputs a clean narrative feels like reality. A pipeline that outputs uncertainty feels like weakness. This is why bad access often masquerades as confidence, and good access often arrives dressed as humility. In a universe where embedded observers only ever receive partial traces, the temptation to worship confident output is everywhere. The Volumetric Time Model doesn't flatter that temptation. It says: if you want to understand your place in time, stop worshipping the story and start interrogating the channel.

The observer is not a magical witness standing in a privileged present. The observer is a physical pipeline, stitched into the world, obeying the same limits as everything else. You see what your channels can deliver. You believe what your channels can support. You act through channels that can be delayed, diluted, and submerged. Your experience of "now" is not a cosmic law; it is the moving edge of your accessible record. Your uncertainty is not proof that reality is being manufactured; it is proof that your record is incomplete. Your certainty is not proof of fate; it is a collapse of plausible histories after a decisive trace finally arrives in a usable form. Once you accept that, the universe becomes less mystical but far more interesting: it becomes a place where knowledge and influence can separate, where legibility can coexist with powerlessness, and where the deepest modern drama is not about what exists but about who has access and who can still couple action into outcomes in time.

To make this vivid, think again about astronomy, because the cosmos is the purest teacher. We can see the birth of stars by detecting infrared light leaking through dust clouds. We can measure the chemical composition of distant atmospheres by studying how starlight is filtered through a thin ring of gas. We can infer invisible mass by how galaxies bend light. The universe is legible because constraints leave signatures. Yet we cannot steer a supernova. We cannot prevent a black hole merger. We cannot reach out and rearrange a galaxy. Our access beam reaches far; our influence beam barely leaves Earth. Nobody finds this morally insulting. Nobody calls it a personal failure. It is simply the geometry of embeddedness and scale. Now recognize that the same geometry exists at smaller scales whenever the relevant levers are delayed, indirect, or drowned in noise. You can have exquisite access and trivial influence. You can watch the corridor tighten and still be unable to move sideways into a safer corridor because the window for meaningful coupling has already narrowed.

That recognition should change how you think about what it means to be "in the present." The present is not a point in spacetime that the universe highlights for you. It is a boundary in your own information structure. On one side: what has arrived through your channels and been stored as record. On the other: what has not yet arrived. The boundary moves because signals arrive, memory accumulates, and your pipeline updates. In that sense, time's arrow is not primarily a metaphysical property you must debate; it is a lived property of record growth. The arrow is the direction in which your accessible evidence tends to accumulate. That is why experience feels directional even in a world described by timeless equations. It is not that equations are wrong. It is that you do not live inside the equations; you live inside a pipeline that receives evidence in sequence.

This also reveals why modern life can feel so strange. A person can be surrounded by forecasts, alerts, dashboards, models, and still feel increasingly helpless. That is not a paradox; it is the natural consequence of incoming channels improving faster than outgoing channels. Sensors get cheap. Data becomes abundant. Models become sophisticated. Meanwhile, the pathways from decision to effect become longer, slower, and more crowded. The more complex the environment, the easier it is for influence to be washed out by competing forces. So the modern observer is often condemned to an uncomfortable posture: you can see farther, yet you can't necessarily touch more. Your pipeline becomes a telescope without a hand. You become a witness with high-resolution regret.

But there is also a kind of liberation in this honesty. Once you accept that access and influence are separate, you stop making the most destructive category error of all: you stop blaming yourself for not steering processes whose coupling scale is beyond you. You stop treating legibility as a promise of control. And you stop demanding metaphysical stories to soothe the pain of powerlessness. Instead, you can ask sharp, practical questions: where does my access come from, how delayed is it, what could fool it, and what windows still exist where action can couple above the noise floor? Those questions are not spiritual. They are physical. They are the right questions for any embedded agent trying to survive in a constrained world.

This is also why the next steps of the book matter. If the observer is a pipeline, then the question of agency becomes measurable. Agency is not a mood. It is not a self-help slogan. It is the existence of a detectable channel from action now to outcomes later. And because channels degrade with delay and noise, there will be a boundary—a horizon—beyond which influence becomes indistinguishable from background. That horizon will not be a metaphor; it will be something you can estimate, test, and calibrate against null worlds. And once you can estimate that horizon, you can define the most disturbing regime: the regime where prediction remains above null while influence is already below detectability. That regime is not a science-fiction plot twist; it is what complex delayed environments naturally produce. It is the regime that makes people feel like they live in a glass cage: they can see everything, but their touches don't register.

Reality is not your feed. Reality is what exists under constraints. Your feed is what arrives through channels. Your belief is your best weighting over which constrained histories remain compatible with the feed. Your feeling of "now" is the moving edge of what has arrived. Your confidence is a property of how narrow your compatible corridor has become. Your disagreement with others is often a property of mismatched pipelines. Your surprise is often a late decisive trace collapsing the story you were leaning on. And your power—your real, physical power—is the ability to couple action into outcomes through channels that can survive delay and noise.

Once you hold that contract, you are ready for the next chapter's hard pivot: away from poetic language about knowing and toward an operational definition of influence. Not what you wish mattered, not what you believe should matter, but what leaves a signature above the noise, at nonzero delay, in a world where shared structure can fake control. That is where the book stops sounding like a philosophy of time and starts sounding like a physics of leverage.

FORMULA DECK (Chapter 3)

The Observer as a Data Pipeline, Sampling, Compression, Null Worlds

1. Underlying history and the observer channel. Let ϕ be an admissible history in the set H on the bounded region V. The observer does not access ϕ directly. Instead, the observer receives observations through a measurement channel. Each observation is generated by a time dependent feature map applied to ϕ, plus noise. The noise may include instrument error, environmental contamination, and unmodeled variability.
2. The accessible record. Define the accessible data up to time t as the record D_t, meaning the collection of all observations received up to that time. This is the concrete record delivered by the pipeline.
3. Information growth as a filtration. Define the information set F_t as the information generated by the record D_t. The filtration property captures monotone growth of access. Later information includes earlier information.
4. Sampling and quantization, why pipelines create artifacts. A practical pipeline samples a continuous signal at discrete times and then quantizes it into finite precision. Aliasing, threshold flips, and apparent pattern changes can be consequences of sampling interval and

quantization rather than changes in the underlying history ϕ.

5. Compression as mapping from record to summaries. Pipelines rarely keep raw records. Define a compressed representation z_t as a summary of the record, produced by smoothing, aggregation, feature extraction, or a learned encoder. The observer effective access is mediated by the compressed representation, not by the raw record. Different compression choices can change what appears real without changing ϕ.
6. Reconstruction and inference layer. Given a prior over admissible histories and a likelihood induced by the observation model, the pipeline forms a posterior distribution over histories compatible with the record. The pipeline world model is therefore a computed object derived from partial access.
7. Posterior concentration as an access metric. Define an uncertainty measure over histories, such as posterior entropy. Define access gain as the reduction in uncertainty when new observations arrive. Large access gain means new evidence meaningfully narrowed the compatible corridor. Small access gain means the pipeline added little usable constraint.
8. Null worlds for checking pipeline made structure. To test whether an apparent dependency is pipeline made, generate null versions of the record that destroy the target

alignment while preserving basic texture. A shuffle null permutes samples within blocks so marginal distributions and local structure remain similar while cross alignment is broken. A shift null circularly time shifts a series by a random offset so internal autocorrelation is preserved while alignment to other streams is broken. If a claimed dependency disappears under these nulls, it is likely spurious or pipeline made. If it survives strong nulls, it is more likely to reflect shared structure or a real coupling.

9. Robustness to preprocessing choices. Let T be a statistic or detector computed from the record or from the compressed representation. Let C_i be a family of reasonable preprocessing maps.

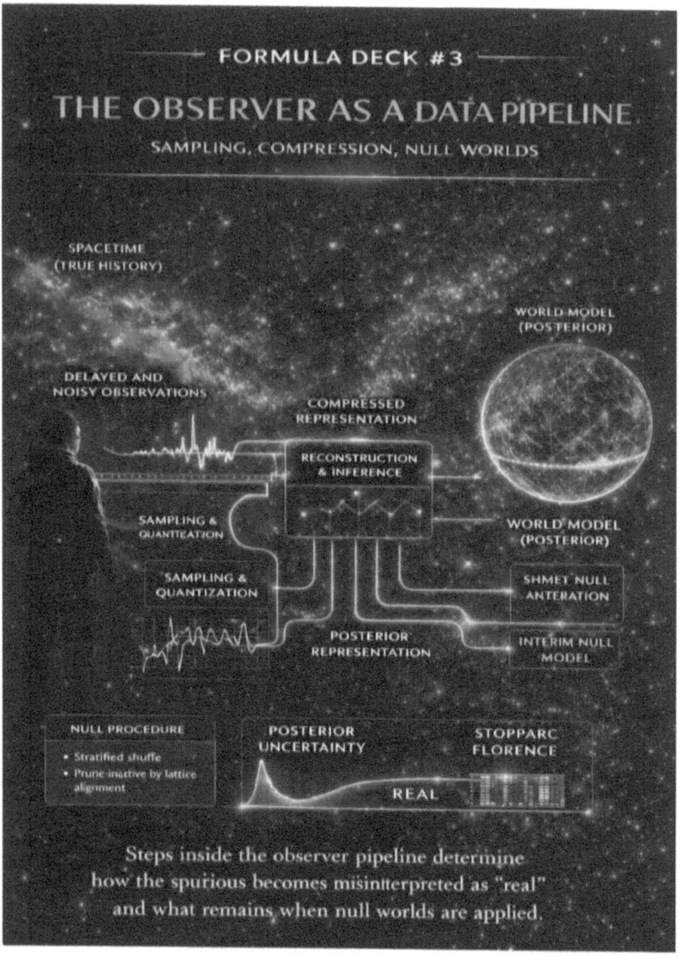

Steps inside the observer pipeline determine how the spurious becomes misinterpreted as "real" and what remains when null worlds are applied.

Chapter 4

Agency Is a Channel

A hand reaches for a control stick.

A real hand. Warm. Slightly sweaty. Nails bitten down to the quick, because the body knows something the mind keeps trying to forget: the world is not where your fingers are. The world is over there. Far. Separated by distance, by delay, by the thin ugliness of transmission.

On the wall: a screen, clean and bright, pretending to be a window. A machine—somewhere else—waits for your touch to become motion. A drone over a coastline. A rover in dust. A crane on a platform swaying in wind you can't feel. The camera feed shows the hook hanging like a question mark above a load that costs more than your entire childhood.

You nudge the stick left.

Nothing happens.

Not immediately. Not in the way your muscles evolved to expect. Your brain is built for tight loops: reach, touch, feel. You are built for near. But here the loop is stretched, and the stretch changes what you are. It turns confident intent into a gamble.

A half-second later the hook begins to drift left—too much. You correct, right. The hook keeps drifting left because your correction hasn't arrived yet. So you correct harder. Now you're fighting the ghost of your own last movement. You're wrestling your previous self across a small dead zone of time.

You can keep doing this—correcting the past with actions that land in the future. You can keep *trying*. But if the loop is delayed enough, your trying becomes its own turbulence. You create oscillations. You create overshoot. You create the very instability you're hired to prevent.

This is the first truth about influence:

It is not the emotion of steering. It is not the moral purity of wanting. It is not the righteousness of responsibility. Influence is the physical fact that an action you take can travel forward through the world's mess and still show up as a difference you can detect—after the world has had time to move.

The second truth is sharper:

Prediction is cheap compared to control.

You can often see what's about to happen long after the ability to stop it has already died.

People confuse these because they *feel* similar. You watch the hook drift. You know it will smash. You know it. Your eyes are correct. Your model is correct. Your dread is correct. And still the load swings into disaster like it never heard you.

That gap—between knowing and changing—is not philosophy. It is geometry plus delay plus noise plus bandwidth. It is a channel.

Think of "agency" as a channel the same way you think of sound traveling through air. Sound is real, but it is not guaranteed. It attenuates. It is masked. It is distorted. It is drowned by other vibrations. You can shout into a storm and still be unheard without anyone being evil. You can shout into vacuum and there is nothing to carry the shape of your voice at all.

Influence is like that.

An action has to be encoded into the world, carried forward, and decoded as a measurable change in later outcomes. The world is not obliged to preserve your signal. The world is busy. The world is loud. The world is full of uncontrolled variables that happily scramble your intent into statistical dust.

So if you want to say "I had agency" in a way that does not collapse under scrutiny, you need a criterion that does not depend on vibes.

Here is the cleanest version in ordinary language:

An agent has influence at a given delay if changing the agent's action changes the distribution of later outcomes, after accounting for what was already knowable at the moment of action.

That last clause matters. It is where most self-deception hides. People declare influence when they merely acted at the same time as an outcome that was already determined by accessible conditions. They confuse being early with being causal. They confuse being present with being powerful.

A channel view forces discipline. It demands you separate three ingredients that get tangled in the mouth:

1. **Signal:** the variation in outcomes that is attributable to action differences.
2. **Noise:** the variation in outcomes that comes from everything else.

3. **Delay:** the time you must wait before the world has had enough freedom to reveal whether your action did anything beyond your immediate measurement pipeline.

Delay is the knife.

Because delay turns a comfortable illusion into a test.

If you measure "effect" at the exact same instant you act, you are standing in a hall of mirrors. Your action and your measurement can share hidden machinery: shared clocks, shared buffers, shared features. You can "cause" a change in the readout without causing a change in the world. The readout is not the world. The readout is a record produced by a pipeline, and pipelines have habits.

So the first rule of honest influence is brutally simple:

If an effect does not survive a positive delay, treat it as suspect.

Not "impossible." Not "false." Suspect.

Because zero-delay coupling is where the cheats live—even when no one is trying to cheat. Even when everyone is sincere. Especially when everyone is sincere, because sincerity makes people stop checking.

There are three classic illusions that fake agency.

The first illusion: shared-state glamour

Imagine a streetlight that turns on when the sky gets dark. You walk outside at dusk and flip a switch inside your house at the same moment the streetlight turns on. You can build a narrative in which your switch controls the streetlight. You can swear it. You can do it three nights in a row. You can bring witnesses.

You are not controlling the streetlight.

Both your switch-flip and the streetlight's activation are responding to the same hidden variable: the falling light level. The world state changed; you and the streetlight both moved in response.

This illusion persists in far more sophisticated places. In markets. In weather. In crowds. In any environment where many actors read the same underlying conditions. You can "predict" a movement, act in alignment with it, and then mistake alignment for leverage.

A channel view says: if your action is just another downstream response to shared conditions, then the coupling you observe is not proof of control. It is proof of correlation driven by common cause.

The second illusion: same-moment contamination

Some measurements are not even trying to be fair about time ordering. They are computed from overlapping data windows. They share features. They reuse the same raw stream. They are "summaries" that blur what happened first and what happened second.

When you act and measure effect in the same moment, you are allowing your action to leak into the measurement in ways that have nothing to do with the world carrying your influence forward. It is like clapping and claiming you moved the echo. You moved the sensor.

This is why "instant success" is so seductive and so dangerous. Instant success often means you changed the display, not the later reality. Instant success often means you found a path around the world's dynamics and into the measurement apparatus.

The channel view demands time separation not as ritual, but as hygiene.

Make the world do work. Make the world carry the signal. If it cannot, then your "agency" is confined to the readout layer.

The third illusion: pipeline hallucination

Even if you avoid shared-state glamour and same-moment contamination, you can still hallucinate influence through the record pipeline itself.

Every observer is forced to sample. Sampling is compression. Compression is loss. Loss is interpretation. Interpretation is bias. Bias is not evil—it is necessary. But it changes what "evidence" means.

Two observers can watch the same unfolding reality and disagree because their pipelines differ: what they sample, how they compress, what they treat as "signal," what they discard as "noise," how they align timestamps, how they smooth, how they label.

A pipeline can create the appearance of stable coupling by enforcing stability: smoothing away volatility, snapping ambiguous events into discrete bins, forcing continuity where the underlying process is jagged.

That can make the world look more steerable than it is. It can also make the world look less steerable than it is. Either way, the pipeline becomes a participant. Not a neutral window.

That is why any serious definition of influence has to be conditional on what was accessible through the record at the time of action. Otherwise you give the observer the power of hindsight and call it agency.

And hindsight is a drug.

Hindsight lets you say: "I could have acted differently," while quietly importing information you did not have at the time. It lets you hallucinate control as an afterimage of regret.

A channel definition rejects that. It says: agency is measured relative to the information you actually carried in your hands, not the information the universe reveals later.

Now take these three illusions—shared-state glamour, same-moment contamination, pipeline hallucination—and notice what they have in common.

They all produce coupling that looks strong at zero delay and evaporates when you force time separation.

That is why the post-zero test is not a preference. It is an ethical line. It is how you stop mistaking mirrors for doors.

Once you accept influence as a channel, a few consequences drop out immediately, like bolts from a loosened frame:

- Increasing delay makes control harder even if prediction stays good.
- Increasing noise makes influence indistinguishable from absence even if effort increases.
- Increasing action bandwidth helps only up to the point where the world's own variability dominates.
- And most painfully: there exists a region where trying and not trying cannot be told apart by any

honest measurement.

Not because the universe hates you. Not because fate is moral. Because the channel is finite, and the world is loud.

The hook keeps drifting. You correct again. You feel the loop slipping. The screen is smooth. Your intent is pure. The world is indifferent to purity. The world only cares whether your action survives transport.

That's the threshold you're approaching: the point where your "control" becomes choreography performed for your own nervous system, while the future continues on its selected path as if your hand were resting in your lap. If you want a boundary you can't argue with, you start here: influence must survive the world's motion. Influence must survive time.

There is a cruel trick the world plays on anyone who wants to feel powerful.

It lets you be right over and over, until correctness starts to feel like agency. You say, *I knew it.* You say, *I called it.* You say, *I moved first.* You say, *I touched the lever.* And the outcome arrives on schedule, obedient, as if the universe is nodding at you.

Then you try the simplest test—change what you do—and nothing changes. The same thing happens anyway. The universe keeps nodding, but now you realize it was never nodding at you. It was nodding at itself.

This is where influence becomes a serious object rather than a mood: you have to stop measuring your actions against the world and start measuring your actions against an alternate version of the world that did not receive them.

That alternate version is not a fantasy. It is the minimum standard for causal honesty.

Imagine two copies of you in two copies of the same room. Same screen, same hook, same wind, same noisy motors, same hidden variables humming under the floorboards. The copies share the same internal record: the same camera feed, the same sensor readings, the same memories up to the instant of choice. Their bodies are even poised in the same tension.

One copy nudges left. The other nudges right.

If later outcomes look different in a way that cannot be explained by chance, then your action has traveled through the world and survived. If later outcomes look the same, then your action did not possess leverage at that delay. It may have been expressive, cathartic, virtuous, dramatic—but it was not operational.

That twin-world image is the cleanest way to understand why "what you knew at the time" matters.

Because when people argue about influence, they constantly smuggle in forbidden information. They look at what happened later and quietly assume they were entitled to that knowledge earlier. They revise the story of their decision using hindsight as if hindsight were part of their senses.

It isn't.

Hindsight is a luxury provided after the fact, like a souvenir. It is not a tool that existed inside the moment.

So any honest definition of influence must be anchored to the record available at the moment of action. Not your future diary. Not the historian's perspective. Not the courtroom reconstruction. The record you actually had—imperfect, delayed, compressed, biased.

This isn't just philosophical tidiness. It changes what counts as agency.

Because a world can be perfectly deterministic and still feel unpredictable to an embedded observer, and a world can be noisy and still feel controllable if feedback is fast enough. Control is not about metaphysical freedom. Control is about accessible difference-making under constraint.

Think about a simple everyday case that exposes the difference immediately: a door with a motion sensor.

You approach. The door slides open. Your brain writes a story: *I triggered it.* That story is correct in one sense—your presence is a condition. But if the sensor was already triggered by the person walking just ahead of you, and the door would have opened anyway, then your presence did not add influence in that moment. You were coincident with the event, not causal in a way that creates a difference.

Now make it sharper.

Suppose the door is faulty and occasionally opens on its own due to electrical noise. You walk up, it opens, you celebrate your magical aura. You walk away, it closes. You do this three times in a row and your confidence hardens into superstition.

The superstition is born from a measurement mistake: you did not compare the world with your action to the world without it under the same accessible conditions. You treated coincidence as proof.

This is the universal failure mode. It scales from doors to geopolitics.

To avoid it, you need a discipline of conditioning, which can be said without symbols:

When you ask whether an action had influence, you must hold fixed what the actor could reasonably know at the moment of choosing.

That fixed knowledge acts like a filter. It tells you which differences are "new" and which differences were already encoded in the moment.

Here's the nasty part: a lot of what looks like influence is just selection.

You choose to act when conditions are favorable. Of course the outcome is favorable. You're not steering; you're timing. Timing can be valuable—sometimes timing is the only thing you have—but timing is not the same as pushing the world into a different trajectory. Timing is surfing a wave you did not generate.

A person stands by a river and waits for a log. When it comes, they jump on it and ride downstream and tell the village they commanded the water.

This is why control claims are so difficult in environments with strong shared signals: markets, crowds, weather-like regimes, anything that moves in patterns many actors can observe. Everyone responds to the same underlying state, and the underlying state does most of the work. If you simply act when your internal cues say "now," your action will be correlated with outcomes even if it has no leverage.

Correlation is cheap because the world gives it away for free.

Control is expensive because it requires your action to survive the world's chaos and still produce a detectable shift in what happens later.

Now take the channel view and add a painful layer: the record itself can be part of the channel.

Your "knowledge at the time" is not a pure mirror; it is a manufactured product. It is sampled, aligned, denoised, compressed, labeled. The record is the result of a pipeline, and pipelines are not neutral.

A pipeline can create apparent influence by making your actions and outcomes share the same upstream processing. This is how many illusions remain intact even when people think they're being careful. They are careful about the world, but careless about the record.

Here's a concrete example in the language of ordinary life.

You have a fitness tracker that predicts you'll sleep poorly tonight. You see the warning and decide to go to bed earlier. The next morning the tracker reports "good sleep." You conclude: *I improved the outcome by acting.*

But if the tracker's sleep score is heavily influenced by when you went to bed—if the scoring algorithm uses bedtime as an input or as a proxy—then your action may have changed the score without changing the underlying physiological reality you care about. You steered the meter, not the body. You gained influence over the record, not over the world the record attempts to describe.

This does not mean your action was useless. It means you must specify what outcome you mean and where it lives. If the outcome is "score," then yes, you had influence. If the outcome is "recovery," the story may be different.

This is the channel view turning into a knife again: it makes you say what you mean.

When you say "I influenced the future," do you mean the future of a label? A display? A reported metric? A sensor's output? Or do you mean the future of the underlying process that the metric claims to represent?

Both are futures. Both can be real. But they are not the same.

And if you can influence the display at zero delay while failing to influence the underlying process at positive delay, then you have discovered the simplest and most common form of counterfeit agency: agency over the readout.

That counterfeit is everywhere. It is the foundation of many modern rituals. The world encourages it because it is cheap. You can "take control" by rearranging dashboards and changing settings and clicking confirmations, and the feeling of control arrives instantly because the readout obeys you instantly.

But influence worth naming must survive the world's own unfolding.

This is why insisting on a positive delay is not pedantry. It forces the distinction between *editing the record* and *editing the world.*

So the discipline looks like this:

- Define what counts as the outcome in a way that

lives outside the immediate readout loop.

- Compare what happens after an action to what would have happened without that action under the same accessible conditions.
- Demand that the difference persists when time separation is enforced.
- Treat any same-moment coupling as suspicious until proven otherwise.

Once you adopt this discipline, you start seeing agency as a scarce resource with a budget.

There is only so much difference-making you can push through a delayed, noisy channel. Past a certain point, your actions are swallowed by the world's variability. Past a certain delay, your corrections arrive too late to couple with the unfolding state in a way that changes what comes next.

And here is the most unsettling consequence, the one people resist because it insults the ego:

You can retain the ability to forecast long after you have lost the ability to steer.

You can watch the river and become extremely good at predicting where the log will go, and still be unable to push the log onto a different branch once it has passed a certain point. You can be a brilliant prophet and a useless controller in the same universe, in the same hour, with the same brain. That split is not a failure of character. It is the channel showing its limits.

There's another way to feel the split between knowing and changing, and it's uglier because it happens in your own body.

Take a shower with a water heater that responds late.

You turn the dial toward hot. Nothing happens. You turn it farther. Still nothing. Your skin starts to resent you. You turn it farther again because you're tired of waiting, and then—finally—heat arrives like revenge. Not a gentle correction. A surge. You jerk the dial back. The heat keeps rising because your correction hasn't arrived yet. Now you're doing that stupid dance: too hot, too cold, too hot again, the loop never settling because the feedback is stale.

You were "in control" the whole time. Your hand was on the dial. Your intention was clear. And yet the outcome behaved like your intention was irrelevant.

That's delay.

Now add noise.

The water pressure wobbles because the building's pipes are shared. Someone flushes. Someone starts the dishwasher. The heater is aging. The sensor is cheap. Your perfect corrections land into a moving target. You can't fix what refuses to be fixed because the environment is injecting variability faster than you can counter it.

This is what it means to say agency is a channel: your action must travel through a medium that distorts it, delays it, and drowns it.

A channel has a capacity. There is a limit to how much distinct influence can pass through per unit time. You can't exceed it by trying harder. Try-hard is not a physics primitive. A channel doesn't negotiate with your sincerity.

That's the core insult: the world does not grade you on effort. It grades you on detectable effect.

So let's say this in the most concrete way possible, without symbols, without ceremony.

At some moment you choose an action. Call it a move, a push, a word, a click, a steering adjustment, a refusal, a yes. Whatever.

Later, something happens. Call it an outcome. A temperature. A collision. A price. A heart rate. A vote. A crash. A door opening. A relationship ending. An alarm firing. A plane landing.

If you want to claim influence, you must be able to say: **had I acted differently, the later outcome would have been measurably different**, given everything that was already visible at the moment of choice.

Not "maybe in a story." Measurably different in a way that survives the world's own turbulence.

This is why agency is not a single property you either have or don't have. It depends on delay. It depends on how far forward you ask your action to matter. It depends on what the world does in the meantime. It depends on what you can actually observe and condition on.

And it depends on something people don't like to admit:

Sometimes your influence is real, but too small to detect without repetition.

Imagine you have a coin that's almost fair. You believe you can bias it slightly by the way you flip it. You can't prove it in one toss. You can't prove it in ten. Your gut will lie to you because your brain is built to see patterns even in noise. If you run ten thousand flips and you're careful, you might see the bias emerge like a faint watermark.

That's not a moral issue. It's statistics.

Influence that exists below the detection floor is operationally similar to no influence, unless you can gather enough evidence fast enough to act on it. If you need ten thousand repetitions to detect your effect, but the situation demands one decision, then your effect is not leverage in that situation. It's a ghost. It might be real, but it cannot be used.

This is another sharp boundary hidden inside the channel view: **usable influence is not the same as influence in principle**.

Usable influence is influence you can detect and exploit within your time and energy constraints. If the world moves faster than your ability to confirm, you are driving in fog. You might still be turning the wheel, but you can't tell if the road is responding or if you're just comforting yourself.

Now bring back the earlier warning: zero-delay coupling is suspect.

Why? Because at zero delay, you can win by cheating the measurement pipeline even accidentally.

There's a reason magicians love close-up tricks. When you watch from too close, you confuse the performer's hand with the object being manipulated. You think the coin moved because you saw a motion. But you didn't see the coin move. You saw a gesture that your brain interpreted as the coin moving. The "effect" happened in the observer.

Zero-delay effects are often like that. They happen in the record. They happen in the scoring rule. They happen in the shared machinery that produces both the action log and the outcome log.

To understand this, picture two clocks that aren't quite aligned. You press a button and the log says you pressed it at 12:00:00.001. The sensor reports the outcome at 12:00:00.001 too. The software rounds both to 12:00:00.000 because it stores time in coarse bins. Now your action and outcome appear simultaneous. You compute a "strong coupling." You celebrate.

But the real world had an order. The real world had delays. Your measurement pipeline destroyed the order and replaced it with a convenient fiction. Your coupling is an artifact of how you wrote time to disk.

Even worse: some outcome variables are computed using inputs that include the action itself. Not maliciously. Just because engineers like efficiency. A control signal is treated as a feature. The "outcome" is a scored quantity that quietly depends on your own input. You can raise the score by pushing the input harder. The score responds instantly. It feels like power.

It is power over the meter.

The meter is not nothing. It has consequences. People live and die by meters. But if you care about the underlying process and not the readout, you need to know which world you're moving.

Here's the most unforgiving way to phrase it:

If your action can alter the reported outcome without passing through the world's dynamics, you have not demonstrated forward influence; you have demonstrated access to the reporting layer.

That's why time separation matters. A positive delay forces the world to "carry" your action forward. It forces your action to survive contact with everything you do not control. It forces the difference-making to be real rather than local to the measurement pipeline.

Now, a person will immediately try to wriggle out by saying: "But some real control is instantaneous."

Sure. In a tightly coupled physical loop—your hand on a cup, your foot on the brake—you can have rapid influence. But even there, the world has delays, and they matter. Neuromuscular lag. Tire friction. Reaction time. Sensor inertia. The world does not permit true simultaneity in the way our stories pretend it does. Zero-delay is a mathematical convenience, not a lived feature.

So when you see coupling at zero delay, you treat it as a diagnostic: it might tell you something is connected, but it doesn't tell you you can steer the future. It's like seeing two wires touching. Contact is not the same as transmission.

Now let's talk about a subtler trap: the shared-state glamour can survive positive delays too.

A crowd is a living example. You shout "Run!" and people run. You feel powerful. But maybe the crowd was already about to run because they saw smoke behind you. Your shout is correlated with running because you're reacting to the same smoke. If you had stayed silent, the running would still happen. Your voice is not the cause; it's another symptom of the shared underlying state.

This becomes brutally confusing in any environment where many agents observe and react: markets, traffic, social dynamics, competitive games. You can be "predictive" because you are reading the same signals, but still have minimal leverage because the outcome is dominated by the shared state and the collective response.

So an honest test of influence requires more than delay; it requires conditioning on what was visible.

When you ask "did my action matter," you don't compare outcomes across all situations. You compare outcomes across situations that were informationally equivalent at the moment of action. You hold fixed what could be known. Otherwise you will attribute to your action what belongs to the visible setup.

This is why agency is not just a channel; it is a channel relative to a record.

Your record is what you carry forward: sensor readings, memories, context, logs, perceptions, summarized and compressed. That record is all you had when you chose. If your action's supposed influence vanishes once you account for the record, then your "power" was just timing within the shared state.

Now look at how these pieces fit together:

- Delay tests whether your action survives transport through the world.
- Conditioning tests whether your action adds anything beyond what was already visible.
- Detection floors test whether the influence is usable within constraints.
- Measurement hygiene tests whether the coupling is real or a pipeline artifact.

When you combine them, you start to see why power is rarer than it feels.

A lot of what we call power is access to controls that change readouts quickly. A lot of what we call power is being early to visible signals. A lot of what we call power is the privilege of acting with better records—faster feedback, cleaner observations, higher bandwidth.

And that last one points toward the next inevitability:

If influence is a channel, then there will be a boundary—depending on delay and noise—beyond which action and inaction cannot be distinguished in their downstream consequences. Past that point, intent becomes theater. Past that point, you can still forecast—sometimes extremely well—but your steering falls below the floor of detectability.

That boundary is not a metaphor. It is a measurable threshold where the channel stops carrying your signal in a usable way.

There is a kind of power that feels like power until you notice it never pushes back on the future. It's the power of reception.

A ruined city has a radio station still broadcasting. The voice is calm, uninterrupted, almost smug. You can hear the forecast. You can hear the warnings. You can hear the countdown of storms and shortages and incoming convoys that never arrive. The signal reaches you because the station is tall and the antenna is clean and the generators are fed.

Now you try to answer.

You press the transmit button and speak your own warning—about the fire spreading through your block, about the bridge collapsing, about the people trapped on the wrong side of the river. You speak into the mic like it matters. You say coordinates. You say names. You say "please." You say "now." Nothing returns.

Not because nobody cares. Not because your words are morally insufficient. Because your transmission channel is weaker than your reception channel. Your battery is dying. Your antenna is bent. Your frequency is drifting. Your message is swallowed by noise and distance and the hard fact that the air is not obliged to carry your intent.

That asymmetry—high-quality observation with low-quality influence—is everywhere, once you learn to see it. It is the quiet architecture underneath a lot of modern despair: you can know more than you can change.

People often treat this as a spiritual problem. It isn't. It's a channel problem.

There are two flows in any embedded life:

- The flow of information into you: what the world allows you to observe and compress and remember.
- The flow of influence out of you: what your actions can actually shift downstream.

Those two flows are not equal. They don't even have to be in the same universe of capability. You can have a world where observation is extremely rich and influence is extremely thin, and you can have a world where influence is thick but observation is crude. Your lived reality—your sense of "freedom," your feeling of "control"—is largely the ratio between those two flows.

This is why it's so common to be correct and still powerless.

A spectator at a chess match can predict the blunder seconds before it happens. They can see the trap. They can feel the inevitable fork. Their prediction channel is active. But their influence channel is zero. Their knowledge is loud in their head and silent in the world.

Now imagine a person sitting at the board with a delay between thought and hand. Their fingers tremble. Their brain is fast but their movement is slow. Their influence channel exists, but delay and noise are eating it. They can see the right move and still fail to transmit it through their body in time. The outcome punishes them the same way it punishes someone who never saw it at all.

This is the core insult of the channel view: the universe does not reward "seeing." It rewards "coupling."

So if you want a definition of agency that survives contact with ego, you need to make it operational in a way that defeats the radio illusion. You need to be able to tell the difference between "I received information about what would happen" and "I sent a signal that changed what happened."

The clean way to do that is not to argue metaphysics. It's to ask a brutal question:

If you swap your action while keeping the moment's accessible record the same, do the later outcomes shift in a way that survives noise and delay?

Everything hangs on two phrases in that sentence: "swap your action" and "keeping the record the same."

Swapping your action is the counterfactual test. It is the smallest demand of causal honesty. If you never imagine doing otherwise, you can't distinguish steering from timing.

Keeping the record the same is what prevents the oldest cheat: using hindsight. You are not allowed to compare yourself-with-action to yourself-without-action if the second version is granted information the first did not have. That's not a counterfactual; it's a different creature.

So the test is: two worlds, same local evidence in the moment, different action, then compare what comes later.

That's the ideal. In reality you can't clone the universe on demand. But you can approximate the ideal in two ways that matter.

One is repetition. When a situation repeats under similar accessible conditions, you can treat each episode as a sample. That's what you do when you flip a slightly biased coin ten thousand times, or when you watch a delayed controller oscillate again and again, or when you test a rule across many runs.

The other is randomization. Randomization is not superstition; it is a tool for breaking shared-state glamour. It is how you stop your action from being entangled with the underlying conditions you're reacting to.

If you decide your action based on what you see, then your action is correlated with what you see, and what you see is correlated with what happens. That triangle can produce beautiful illusions. You can look powerful without moving anything.

Randomization breaks the triangle.

You flip a coin to decide whether you act. Not because you are surrendering agency, but because you are isolating it. You are forcing your action choice to be independent of the visible state, so that if outcomes still differ downstream, you have evidence that your action carried influence through the world rather than merely riding the wave of shared conditions.

This is the difference between "I bought before the price rose" and "my buying moved the price." One may be true while the other is false, and the only way to tell is to separate your decision from the cues that would have made the outcome happen anyway.

Now add the second knife: delay.

Even when randomization is perfect, you can still fool yourself if you evaluate influence too close to the moment of action. Because the record layer can be contaminated. The readout can respond instantly to the input in ways that do not represent real downstream change.

So the discipline is layered:

- Break shared-state glamour with randomization or careful conditioning.
- Break readout contamination by enforcing time separation.
- Break pipeline hallucination by auditing what the record actually measures.

Time separation is what forces the world to become the medium. It forces your action to be carried forward, distorted, attacked by noise, and then possibly recovered as a difference.

That is why a positive delay is the minimum toll you pay to be taken seriously.

And this is also why the zero-delay spike is such a seductive lie. It is the place where everything is easiest to fake. It is where the meter is most likely to be touched by the same hand that moves the knob. You can generate a beautiful coupling at the same instant because the same machinery is producing the action log and the outcome log, and the machinery is not obligated to respect your story about cause and effect.

A positive delay doesn't solve everything, but it forces the problem into the world rather than leaving it inside your measurement pipeline.

Now, when you start thinking like this, a strange reversal happens in how you interpret "control."

Most people imagine control as an absolute. Either you have it or you don't. Either you're free or you're trapped. That binary is comforting because it matches the way blame works in ordinary speech.

But in a channel view, control becomes a curve. It depends on how far forward you're asking your action to matter. It depends on delay. It depends on noise. It depends on how much bandwidth you can push into the world. It depends on how quickly the world returns feedback and how clean that feedback is.

A small delay can be survivable. A longer delay can produce oscillation. A longer still can make your actions indistinguishable from random twitching. Somewhere along that curve, you pass a line where you can no longer demonstrate that you are doing anything at all—at least not within a usable time window. And here's the nasty part: your prediction curve can remain high while your influence curve falls.

You can become exquisitely good at forecasting a collapse and still be unable to stop it. You can watch the temperature rise, the load swing, the crowd panic, the price cascade, the relationship decay, and you can be right about it all. The world will even reward you for being right with the warm narcotic of recognition—*you called it*—while refusing to grant you any actual leverage.

If you don't carry the channel concept, you will mistake that recognition for power. You will confuse the applause of correctness with the physics of coupling.

So you need a measurement posture that is as hostile to your ego as it is to your enemies.

You need "null worlds."

A null world is not the absence of meaning. It is a baseline where any apparent coupling is presumed to be counterfeit until it survives specific tests. You compare your observed influence against versions of the world where the action is deliberately detached from the outcome in ways that preserve everything else: the same rhythms, the same inertia, the same slow drift, the same autocorrelations. If your influence measure stays high under null manipulation, then your measure is not measuring influence; it is measuring structure in the record or shared-state drift.

This is how you keep yourself from worshiping artifacts.

Because the world is full of structure. Structure will happily impersonate agency. If your metric is naive, it will light up even when nothing is being steered. It will tell you you are powerful because the world is patterned.

The correct response is not despair. The correct response is precision.

When you treat agency as a channel, you stop asking "am I free?" and start asking "at what delays, under what noise, with what feedback quality, do my actions remain distinguishable from null?" That question is not existential. It is testable. It produces curves instead of slogans. It produces thresholds instead of myths.

And once you have thresholds, you can talk about boundaries that are not moral, not emotional, not negotiated: boundaries where the channel stops carrying you.

That boundary is coming.

A lot of people think the hardest part of influence is acting.

It isn't.

The hardest part is proving—without lying to yourself—that anything you did actually traveled forward and changed what happened later, rather than merely appearing alongside what was already on the way.

Because the world is not random. The world has inertia. The world has drift. The world repeats itself in lazy, expensive patterns. It carries momentum the way oceans carry heat: stubbornly, silently, for far longer than your attention span. And that structured persistence is the perfect mask for counterfeit agency.

If you measure "effect" using any naive statistic, you will find effects everywhere. You will discover "control" in places where no controller exists. You will find coupling between actions and outcomes simply because both are embedded in a stream that remembers its own recent past.

Here's the simple trap:

If outcomes tend to resemble their own recent values, then almost anything aligned with recent values will appear "predictive," and almost anything timed to recent values will appear "influential," even if the action has zero leverage. This is not fraud. This is autocorrelation—the world's tendency to continue being itself.

People are pattern hunters. Autocorrelation feeds that hunger like sugar. It makes you feel like you're doing something because the future looks like the present, and you were present.

So the first thing you need—before you ever declare agency—is a baseline that humiliates your metric.

You need to ask: *If there were no real influence, would my method still report influence anyway?*

That baseline is a null.

A null is not an insult. It is an honesty device. It is how you force your definition of influence to earn its name.

Take a concrete scenario. You're controlling something remote: a camera on a pole, a rover in dust, a boat you can't touch, a digital avatar in a game with delay. You send an action. You see an outcome. If you just look at the data and ask, "Do action signals correlate with outcomes?" you will almost always find a correlation, because both are reacting to the same unfolding conditions. The environment pushes the machine; you respond; your responses reflect what you saw; what you saw reflects what was already happening. A triangle closes. It looks like power.

A good null breaks the triangle.

One way is to deliberately scramble the relationship between action and outcome while preserving the overall shape of both streams. You keep the rhythm of actions—how often you act, how bursty you are, how your behavior clusters—yet detach which action goes with which later outcome. If your influence measure still lights up after that detachment, your measure is not measuring influence. It is measuring structure that exists even when no leverage exists.

Another way is to shift time. Not by randomizing everything into static noise, but by sliding the action stream forward or backward so it keeps its internal continuity while losing the correct alignment to outcomes. This is a cruel null because it preserves everything that makes the data look real: slow drifts, recurring cycles, bursts, fatigue effects. If your "agency" survives time-shifting, then what you found was not agency. It was a shared rhythm.

Nulls are what keep you from worshiping the world's laziness.

Because the world repeats itself. It does it so reliably that people build religions on it. The sun rises, the seasons change, the crowd panics the same way, the price chart grows familiar shapes, the body responds in habits. You can predict those patterns without influencing them. You can time your actions to them and feel like you're steering.

A real influence measure must fail under a null that preserves the world's inertia.

This brings us back to delay.

Delay is not just a practical annoyance. It's the filter that forces influence to pass through the world rather than staying trapped inside the measurement layer. When you demand a positive delay, you are saying: *I am not impressed by instant compliance from a display. I want to see my action survive the medium.*

But even with delay, you can fool yourself if you don't calibrate.

Calibration is where you stop talking like a philosopher and start behaving like a cautious engineer. You don't say, "This number looks big, therefore I had leverage." You say, "How big does this number get in worlds where I know leverage is absent?"

That is a different posture. It's colder. It hurts the ego. It also prevents you from becoming a cult leader of your own metric.

In practice, calibration looks like this:

You generate a family of null worlds—worlds that preserve everything except the causal link you're claiming. You measure your supposed influence inside those null worlds. That gives you a distribution of "fake influence" values: the baseline amount of coupling your metric reports when it shouldn't.

Then you compare the real world to that baseline.

Not with gut feeling. With a threshold.

You pick a high bar: for example, you might demand that your measured coupling exceeds what appears in 99% of null worlds. That means you are tolerating a small false-alarm rate. You are admitting: "Sometimes the world will fool me, and I will refuse to pretend otherwise."

This is what converts the language of agency from mythology into instrumentation.

It also reveals a deeply uncomfortable fact: agency is not binary, but detectability is.

You can have tiny real influence that exists in principle, and still fail to clear the detection floor. You can have genuine leverage that is too weak to distinguish from null given your sample size, your noise level, your feedback delay, your bandwidth. In that case, you may be "influencing" in the metaphysical sense, but you are not influencing in a usable sense.

Usable influence means: influence you can detect, trust, and exploit within the constraints of time and energy.

A mosquito can shift the trajectory of a truck by a microscopic amount if it hits the windshield at the right moment. That is influence in principle. It is also irrelevant influence. It does not produce a controllable difference you can harness. It does not give the mosquito steering rights.

So you need to be precise about what kind of agency you are defining. The definition that matters for lived life is the usable kind: detectable coupling that survives the world's noise and delay and your own limited sampling.

Now, when you combine delay with calibration, a simple picture starts to emerge—one that will become impossible to unsee.

Imagine increasing the delay between action and outcome, step by step.

At very small delays, your influence measure may be high, but you distrust it because you know the measurement layer is full of shortcuts and shared machinery. Instant coupling is easy to counterfeit.

As you move to slightly larger delays, two things happen:

1. Any "influence" that was actually meter-twisting begins to evaporate, because the world has time to move independently.
2. Any real downstream effect has a chance to reveal itself, because your action must now pass through the environment rather than directly into the readout.

So you watch what your influence measure does as delay increases.

In many real scenarios, it drops. Sometimes it drops smoothly. Sometimes it falls off a cliff. Either way, it tends to approach a floor where it becomes statistically indistinguishable from what your null worlds produce.

That floor is the place where the channel stops carrying your signal in a way you can detect.

This is the birthplace of a boundary that has nothing to do with morality and everything to do with constraints.

You don't have to call it fate. You don't have to call it doom. You can call it what it is: a detectability limit for action-outcome coupling under delay.

And here's the split again, now sharper.

While your influence measure collapses toward the null floor, your prediction ability may remain stubbornly nonzero. You can often continue to forecast because forecasting is mostly about extracting structure from the world's inertia, which can persist across large delays. The world continues to carry its own memory forward, and you can learn that memory without steering it.

So you get the strange modern condition: the ability to see far without the ability to touch far.

This is not a contradiction. It is exactly what you should expect when observation and influence have different channel capacities. The world can be legible without being steerable at the same range. Now notice what calibration protects you from.

Without calibration, you might stare at a small, noisy influence signal at a long delay and declare: "I still have control." You might confuse a faint nonzero value with usable leverage. You might build a story around it, because stories are cheap and comfort is addictive.

With calibration, you are forced to ask: "Is this value above what I see when there is no leverage?" If it isn't, then the honest conclusion is not "I am powerless in my soul." The honest conclusion is: "At this delay, under these conditions, my attempted influence is not distinguishable from null." That is a statement about measurement and constraints. It is also a statement about how far your agency reaches.

And once you can say that sentence, you are one step away from naming the boundary itself: the first delay beyond which influence drops below the calibrated floor.

That boundary is not a slogan. It is a threshold you can estimate, test, stress, and move by changing noise, bandwidth, and feedback quality.

In other words: it is a boundary you can do science on.

Even if the world were perfectly fair. Even if your measurements were immaculate. Even if you calibrated against a ruthless null and enforced time separation and refused to touch the readout layer like a cheap magician.

Even then, influence would still be limited by something simpler than philosophy:

You can only push so much distinct intent into the world per unit time.

Call it bandwidth. Call it action capacity. Call it the rate at which you can produce interventions that are meaningfully different from one another. A hand can only twitch so many times. A voice can only issue so many commands. A controller can only change so many parameters. An organism can only burn so much energy. A mind can only pay attention to so many degrees of freedom without collapsing into frantic randomness.

This is the part people hate, because it makes power boring.

Power becomes not a crown but a budget.

And budgets have ceilings.

Imagine a boat in rough water. The current is strong. The wind changes direction. Waves slap the hull at irregular angles. The boat is not a smooth object in a clean fluid; it's a messy shape in a messy medium. Now give yourself a paddle. You can push. The paddle is real. It moves water. It will move the boat—sometimes.

But if the current is strong enough and the wave noise is high enough, your paddle strokes stop being distinguishable from the random motion the water already imposes. You are paddling inside turbulence. Your effort is real, and yet it becomes statistically invisible. Not invisible to your muscles, but invisible to the boat's downstream position once you account for everything else happening.

Now make the cruel move: shorten the time window. You don't care about where the boat is in an hour. You care about where it is in ten seconds because there's a rock. You paddle. But the boat has inertia, the water has inertia, and the feedback you're using—your eyes, your balance, your reading of waves—is delayed. You are always correcting the last wave.

In that setting, the paddle becomes a symbol more than a tool. You can still say you are "doing something." You can still feel the virtue of effort. But the coupling between your strokes and the boat's actual avoidance of the rock becomes fragile. Sometimes you save yourself. Sometimes you don't. Sometimes you are saved by the water despite your strokes. Sometimes your strokes make it worse by steering into oscillation.

This isn't tragedy. This is what it means for influence to be mediated by a finite channel.

A channel has two limits at once:

- **A noise floor**: below a certain signal-to-variability ratio, you cannot reliably detect your own effect.
- **A rate limit**: above a certain intervention frequency, you inject instability faster than you correct it.

People talk about "taking control" as if it means acting more, acting faster, acting harder. The channel view says the opposite is often true. Past a point, more action is just more noise with your signature on it.

That's why the delayed shower dial creates oscillation: you "act" too frequently relative to the lag. Your corrections pile up and collide. The loop becomes a fight between you-now and you-one-second-ago, both sincere, both wrong in the moment they land.

So a sane controller does something that feels like weakness: it waits.

Waiting isn't resignation; it is timing your bandwidth to the channel's capacity. It is refusing to overdrive the medium. It is choosing fewer, cleaner interventions and letting the world carry them forward long enough to evaluate.

This is where influence stops looking like dominance and starts looking like patience under constraint. Now add the twist that makes the split between forecasting and steering feel inevitable rather than tragic. Forecasting is mostly about extracting the world's structure—its inertia, its slow modes, its repeated rhythms. Those rhythms can persist far into the future because the world is not reinventing itself every millisecond. It drifts. It remembers. It carries momentum across time.

Steering, by contrast, requires you to inject enough distinct intent fast enough to push the world onto a measurably different downstream path. That requires bandwidth and low latency and sufficient energy. It requires a channel that can carry your signal forward without shredding it.

So it is entirely ordinary—almost guaranteed—that prediction remains possible long after influence has fallen below the usable floor.

You can read the waves and know where the boat will drift, because the current is strong and consistent. You can predict the drift better and better as you learn the river. But your paddle, no matter how heroic, may not be strong enough to create a measurable deviation from the drift within the time window that matters.

That's the condition that breeds the deepest confusion: you become more knowledgeable and less effective at the same time.

The naive mind experiences this as betrayal. "If I can see it, I should be able to stop it." That sentence is a moral demand disguised as physics. The world doesn't honor it.

The world says: seeing is one channel; changing is another.

They can decouple.

And when they decouple, you get the strange landscape where warnings become abundant and leverage becomes scarce. You get a culture that produces infinite forecasts and increasingly thin interventions. You get people who can describe collapse with exquisite precision and still fail to move the needle by even one calibrated unit.

Now, to make this more than poetry, you need a way to talk about "usable" influence without pretending you are God.

Usable influence is not the presence of some metaphysical power. It is the ability to produce downstream differences that clear a calibrated detectability bar.

That bar exists because you are embedded. You have finite time. You have finite samples. You have finite energy. You can't run a million repeated trials of your life just to confirm a microscopic effect. You have to act under uncertainty.

So you set a rule that is not elegant but is honest:

If the coupling between your action and later outcomes does not exceed what you routinely see in null worlds, then you do not claim leverage at that delay.

That rule does not say you have no influence in the universe. It says you have no influence you can demonstrate and exploit under your constraints.

This is what turns the channel view from fatalism into a measurement discipline. You stop arguing about freedom in the abstract and start mapping where influence is operational.

Map it how?

By walking the delay forward.

At each positive delay, you ask: when I vary my action, holding fixed what was visible at the moment of choice, does the distribution of outcomes shift in a way that clears the null bar?

At small delays, you might see strong shifts—though you distrust them until you've audited the readout layer. At moderate delays, some effects survive and some die. At larger delays, fewer effects survive. Eventually, you reach a region where any measured coupling is indistinguishable from what your null worlds produce.

That region is not "doom." It is the channel's natural boundary.

And here is the most useful thing about treating agency this way:

The boundary can move.

Not by wishing. By changing the channel conditions.

- Reduce noise: cleaner feedback, fewer uncontrolled perturbations, less environmental turbulence.
- Reduce delay: faster sensing, tighter loops, less stale correction.
- Increase effective bandwidth: interventions that matter more per action, not more actions per second.
- Improve conditioning: better record quality so you don't waste action fighting ghosts created by your own pipeline.

Notice how unromantic this is. It doesn't say "be brave." It says "shorten latency." It doesn't say "be pure." It says "audit your measurement." It doesn't say "try harder." It says "stop overdriving the loop."

That's why this view is so hard to accept: it demotes virtue from the driver's seat. It replaces moral drama with constraint management.

But it also offers something most stories of power don't:

It gives you a way to separate real influence from theatrical motion.

Because theatrical motion is abundant. It is the easiest thing in the world to generate. You can act constantly. You can click constantly. You can react constantly. You can flood the world with micro-interventions until the world is mostly reacting to your noise.

That feels like being alive. It feels like participation. It can even look like "engagement."

But the channel view asks a colder question:

Does any of this survive forward transport through delay and noise in a way that produces calibrated downstream difference?

If not, you are not steering. You are vibrating.

That is not a moral condemnation. It is a description of a coupling that fell below threshold.

And once you accept that, the next step is unavoidable.

If there is a threshold—if there is a calibrated floor under which action-outcome coupling becomes indistinguishable from null—then you can define a boundary in delay space: the first delay at which your influence measure can no longer clear the bar.

Not a poetic boundary. A measurable one.

A boundary where the channel stops carrying you.

And beyond that boundary, something eerie becomes normal: you may remain an excellent forecaster. You may still see the rock. You may still know the boat will strike. You may even know exactly when.

But your paddle strokes—no matter how sincere—become statistically equivalent to silence.

That is the moment you stop confusing correctness with control.

That is the moment you stop mistaking reception for power.

That is the moment you finally admit what the world has been showing you the entire time: influence is a channel, and channels have limits.

There is a moment—quiet, almost polite—when the world stops arguing with you.

Not dramatically. No thunder, no cinematic snap, no moral verdict. Just a soft flattening of consequence. You do a thing, and the future behaves as if you didn't.

Most people never notice it because they keep moving. They keep clicking. They keep speaking. They keep "correcting." They keep feeding the loop with motion, because motion feels like agency even when the channel is already full and the medium is already deaf.

But if you slow down and watch with the cruelty of a scientist and the tenderness of someone who is tired of lying to themselves, you can see the transition.

It looks like this: At first, when you vary what you do, the later outcome varies too. Not perfectly—never perfectly—but enough that the difference survives the world's noise. You can detect it. You can rely on it. You can build a strategy around it.

Then you push farther. You ask your action to matter later. You demand influence across more delay, more intervening chaos, more opportunities for the environment to scramble your intent. You keep measuring. You keep comparing to null baselines that preserve the world's laziness but break any real action→outcome link.

And the effect shrinks.

Sometimes it shrinks smoothly, like a signal fading with distance. Sometimes it collapses abruptly, like stepping off a ledge. Either way, it trends toward a region where your measured coupling is indistinguishable from what you get when you deliberately detach action from outcome.

That region is the graveyard of theatrical motion.

Not because you stopped trying, but because trying stopped being legible to the future.

This is the point where people reach for metaphysics—fate, destiny, meaning, conspiracy, punishment—because metaphysics is emotionally generous. It tells you there's a story behind your loss of leverage. It tells you the universe is talking to you.

But the more honest story is colder and simpler:

A channel has limits. Past those limits, your signal becomes unresolvable from the noise it swims in.

To see why this isn't a philosophical punchline, imagine you're trying to communicate with a flashlight across a foggy field at night.

Close range: you can blink patterns, and the other person reads them. You can send instructions. You can coordinate. Your light has leverage because it produces detectable differences in what the other person does next.

Now add distance. The fog thickens. The field swallows the beam. The other person sees a dim blur, sometimes, maybe, indistinct from the background shimmer.

You can still blink. You can still feel your finger moving. You can still believe you're "sending." But if the receiver can't distinguish your pattern from the fog's own fluctuations, your message is operationally absent. Not morally absent. Not spiritually absent. Operationally absent.

That's what it means for influence to fall below a detection floor.

And now take the image back into lived life.

Your actions are the blinks. The world is the fog.

The fog is delay. The fog is noise. The fog is all the uncontrolled degrees of freedom that move while your intent is trying to travel forward. The fog is also your own limited bandwidth: you can't blink arbitrarily fast without turning your signal into a flicker indistinguishable from random shimmer.

The most painful realization is that you can still observe clearly while your influence becomes fog-bound.

You can still see the other person. You can still see their silhouette against the horizon. You can still predict where they'll walk because the terrain slopes and the path is obvious and humans are pattern creatures.

Your prediction channel remains functional.

Your influence channel dies.

That is the split that turns "being right" into a kind of torture: you know what will happen, but your ability to transmit a difference into what happens is gone.

Once you accept this split, you stop using the word "agency" like a compliment and start using it like a measurement.

Agency becomes a claim with a price.

The price is this discipline:

1. **Time separation.** You don't accept same-moment effects as evidence of forward influence. You treat them as suspicious until proven clean. You force your action to pass through the world's dynamics rather than directly into a display.
2. **Conditioning on accessible evidence.** You don't credit yourself for outcomes that were already baked in given what was visible at the moment. You refuse the drug of hindsight. You ask whether your action added anything beyond the conditions you already had in hand.
3. **Null baselines that preserve structure.** You don't compare reality to a cartoon of randomness. You compare reality to null worlds that keep the world's inertia intact—its drift, its repetition, its slow modes—while breaking the specific action→outcome link you're claiming. If

your "agency" survives null manipulation, then you didn't measure agency; you measured structure.

4. **A calibrated threshold.** You don't declare leverage because a number looks nonzero. You declare leverage only when it clears a bar that your null worlds rarely exceed. You admit false alarms exist and you refuse to pretend otherwise.

When you do these four things, something quietly miraculous happens:

The feeling of agency becomes less important than the mapping of it.

You stop asking "do I have control?" and start asking "how far does my control reach under these conditions?"

That question produces a landscape rather than a slogan.

In that landscape, the boundary shows up naturally. Not as a cosmic law carved into stone, but as an operational limit: the first delay beyond which your action cannot be distinguished from null influence at the chosen confidence level.

Call it a horizon if you want. The word fits because it behaves like one: you can move toward it, you can sometimes push it back by improving conditions, but you can't stare past it and pretend you're steering what you can't even touch.

This is where waiting returns—not as a romantic posture, but as a technical choice.

In a delayed loop, acting too often is not more control; it's more self-interference. You inject new corrections before the old ones have had time to land. Your actions pile up in transit and collide in the future like badly timed waves. The result is oscillation and overshoot: you become the source of instability.

Waiting is how you keep your signal coherent enough to survive transport.

Waiting is how you avoid turning your intent into noise.

And waiting is also how you learn whether you have leverage at all. If you can't tolerate the delay required to see your action's downstream effect, you will forever confuse immediate feedback with real influence.

The world rewards impatience with instant readouts. It rewards constant motion with the illusion of engagement. It tempts you into believing that because something responds quickly, it must be the world responding.

But the world is not the readout.

The world is the medium your action has to cross.

So there is an ethic that falls out of pure measurement, with no sermons required:

Don't congratulate yourself for touching knobs. Congratulate yourself only when the future behaves differently because you did.

And if you want to take that ethic seriously, you have to be willing to accept the result when it says: here, at this delay, under this noise, your trying and not trying are indistinguishable.

That sentence stings because it sounds like a judgment. It isn't. It's a coordinate.

It tells you where you are in the landscape.

And once you have coordinates, you can stop arguing with the fog and start changing what can actually be changed: feedback quality, delay, bandwidth efficiency, and the cleanliness of your measurements.

Now the final turn—the one that makes the entire view feel inevitable:

When influence collapses beyond the boundary, the world does not go blank. The world remains structured. The world keeps drifting in legible ways. Your prediction can remain nontrivial because prediction often rides on the world's inertia, and inertia persists.

So you can still see. You can still forecast. You can still be correct.

But correctness alone is not power. Correctness is reception.

Power is transmission that survives.

Once you accept that, you stop treating "agency" as a moral property and start treating it as a fragile coupling that must be demonstrated, delay by delay, against the fog and against your own appetite for flattering stories.

And you get a final principle that is harsh enough to be useful:

Existence does not guarantee access.

Access does not guarantee influence.

And influence, when it exists, is never infinite—it is a channel with a reach, a budget, and a boundary. Beyond that boundary, your motion becomes choreography, your certainty becomes commentary, and the future continues on its path with the calm indifference of weather—patterned, predictable, and untouched by your blinking flashlight.

FORMULA DECK (Chapter 4)
Agency as a Channel, Post Zero Influence

1. Action and outcome variables. Work inside the same VTM setting, spacetime as (M, g), a bounded region V inside M with boundary ∂V, and an admissible history set H on V. Let an action mean what the agent chooses at decision time t. Let an outcome mean any downstream variable evaluated after a delay, such as a later measurement, an event indicator, or a state dependent quantity.
2. Accessible record and information constraint. Let the accessible record up to time t be the data the agent actually had, measurements, observations, logs, and signals. Let the information available at time t be the information generated by that record. Any claim about influence must be evaluated relative to what was accessible at the moment of choice, not with hindsight.
3. Operational definition of influence as a distribution shift. Agency at a given delay means that changing the action changes the conditional distribution of later outcomes when the informational context is held fixed. In plain terms, different actions, same available information, measurably different downstream outcome distribution. This is the counterfactual requirement in operational form.

4. Quantifying influence as delayed conditional coupling. Define a delayed coupling measure that captures action to outcome dependence after accounting for what was already accessible at decision time. Conditional mutual information is one suitable choice, but the core idea is broader. Influence is measured as the part of action outcome coupling that remains after removing what access alone already implies.
5. Post zero restriction for measurement hygiene. Treat same moment coupling as suspect by default, because it can be contaminated by shared features, logging order, overlapping windows, or readout layer leakage. Operational agency claims are therefore made using post zero delays, meaning delays greater than zero, unless a domain specific justification is provided.
6. Shared state artifact caveat, why conditioning and delay both matter. If actions and later outcomes share a common driver that is hidden or only partially observed, then naive dependence measures can be nonzero even when action has no real leverage. Conditioning on what was already accessible removes dependence explainable by the record. Enforcing post zero delays reduces same moment contamination. Together they target the two dominant counterfeit sources, shared state artifacts and same tick leakage.
7. Null baseline preview, calibrated in the next

deck. Define a null procedure that breaks action to outcome linkage while preserving relevant structure, for example action shuffling within information strata or autocorrelation preserving time shifts. Compute the same steering measure under the null procedure to obtain a null baseline curve over delays. Interpretation rule, a measured post zero steering curve is treated as evidence of influence only when it rises above the null baseline at a chosen false alarm tolerance. Formal thresholding and the horizon definition are handled in the next deck.

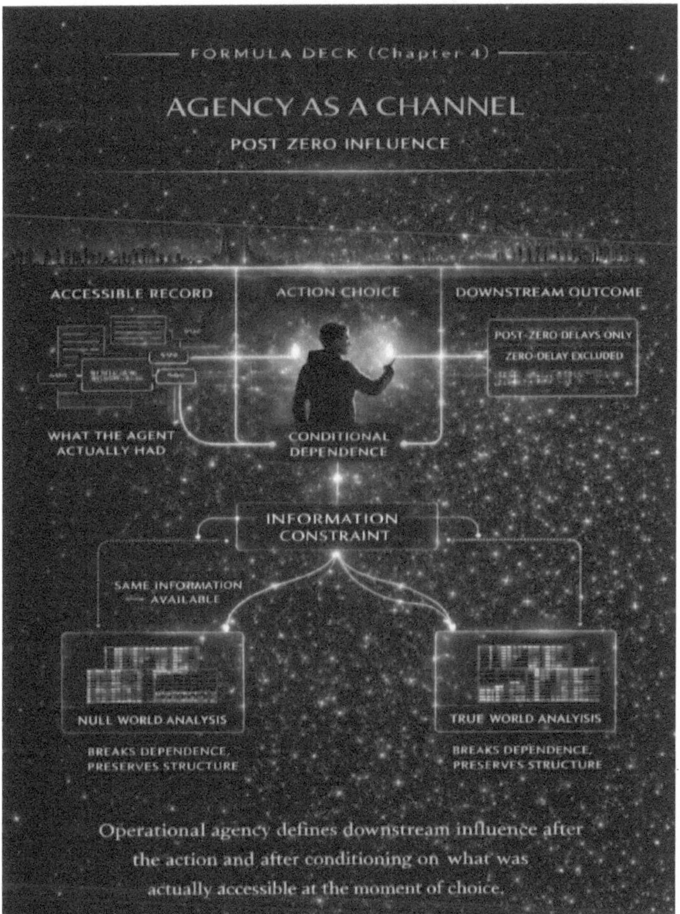
FORMULA DECK (Chapter 4)
AGENCY AS A CHANNEL
POST ZERO INFLUENCE
ACCESSIBLE RECORD
ACTION CHOICE
DOWNSTREAM OUTCOME
POST-ZERO DELAYS ONLY
ZERO-DELAY EXCLUDED
WHAT THE AGENT ACTUALLY HAD
CONDITIONAL DEPENDENCE
INFORMATION CONSTRAINT
SAME INFORMATION AVAILABLE
NULL WORLD ANALYSIS
TRUE WORLD ANALYSIS
BREAKS DEPENDENCE, PRESERVES STRUCTURE
BREAKS DEPENDENCE, PRESERVES STRUCTURE
Operational agency defines downstream influence after the action and after conditioning on what was actually accessible at the moment of choice.

Chapter 5

The Agency Horizon

The fog isn't dramatic. That's the first insult.

It doesn't roll in like a villain. It doesn't roar. It just thickens quietly, until the distance between "there" and "here" becomes a medium you can't argue with. You can still see a little. You can still make shapes out of the darkness. You can still tell yourself stories.

And you can still blink your flashlight.

Once. Twice. Three times. A pattern. A signal. A confession.

Across the field, somewhere beyond the gray curtain, you know there's a person. You can't see them clearly, but you've seen them before. You've watched their silhouette move along the ridge at dusk. You've learned their habits. You can predict the slow drift of their path the way you can predict the slow drift of weather: not because you control it, but because it repeats.

So you blink the pattern again.

And the fog does what fog does: it scatters the beam into a soft bloom, a pale bruise on air. Your careful sequence becomes a smear. The difference between three quick blinks and one long one dissolves. The light arrives as a vague shimmer indistinguishable from the background glitter of moisture.

You keep blinking, because blinking feels like doing something.

You watch for a response. A return signal. A movement. Any proof that your pattern survived the crossing.

Nothing.

Not because the other person hates you. Not because they refused. Not because your message wasn't important. The medium is simply not carrying the fine structure of what you're trying to transmit. The medium is turning your intention into noise.

That's a boundary. Not a philosophical one. An operational one.

It isn't a line painted on the ground. It's a threshold where the world stops being able to tell the difference between your signal and the fog's own shimmer. At that point, your message is not "ignored." It is unreadable.

And unreadable influence is indistinguishable from no influence at all—at least for any creature that has to act under time and energy constraints.

This is the core humiliation that follows anyone who confuses effort with effect:

You can still do the action. You can still spend the energy. You can still feel the virtue of trying. But the future—the part of the world you were aiming to change—stops registering you in a way you can demonstrate.

Most people avoid this realization by never measuring it properly. They keep their evaluation close to the moment, where the beam looks bright in their own hands. They judge success by the feeling of transmission rather than the fact of reception. They declare victory because the flashlight obeyed them instantly, because their fingers moved, because the device clicked.

But the device clicking is not the world changing.

The world changing is the only thing that matters if you're talking about influence.

So the question becomes: how do you locate the boundary without lying?

How do you distinguish "my message didn't get through" from "my message got through but the person chose not to answer" from "my message was never a message at all, just a light flicker swallowed by fog"?

The answer is cruel but clean: you compare your world to worlds where you know influence is broken, while keeping everything else as real as possible.

People resist this because they want a simple test: "Is the coupling nonzero?" They want a number that feels like a guarantee. They want a talisman. They want an easy certificate of power.

But the world is structured, and structure impersonates power.

A river has a current. If you throw a leaf in and it drifts downstream, you can pretend your throw "caused" the downstream motion. You can repeat the throw and the same thing happens, and repetition will harden your superstition into certainty. The leaf's motion is real, your arm is real, and still your claim is counterfeit. The current would have carried it anyway.

This counterfeit becomes more seductive in places where the world remembers itself. Inertia. Drift. Autocorrelation. Cycles. Regimes that persist. When the future tends to resemble the present, almost any action timed to the present will appear "effective," because the outcome was already leaning that way. Correct timing becomes indistinguishable from steering, unless you force the world into a harsher test.

Fog is the harsher test.

Delay is fog. Noise is fog. And structured drift is the most dangerous fog of all because it looks like a road.

So you need a baseline that humiliates your measurement.

You need to ask: If I had no leverage at all, what would my influence score look like anyway, just from the world's own structure?

This is where "null worlds" enter—not as abstract philosophy, but as a survival tool.

A null world is a version of your data where the link you claim—between what you did and what happened later—is deliberately severed, while everything else that makes the world real is kept intact: the rhythms, the bursts, the slow drifts, the tendency of today to resemble yesterday, the stickiness of regimes, the cheap predictability that comes from inertia.

If your supposed influence remains strong inside a null world, then you didn't measure influence. You measured structure.

Structure is abundant. It's the world being lazy. It's the world being expensive to move. It is the reason you can forecast storms and still fail to steer them. It is the reason you can predict traffic and still fail to change it by honking.

So the baseline must preserve structure, otherwise you cheat in the opposite direction: you compare reality to a cartoon of randomness and declare yourself powerful because reality is patterned. That's like comparing a real river to a painted river and concluding your paddle works because the painted river never moves.

A serious baseline keeps the river real and removes only your paddle's downstream effect.

Then—and only then—do you earn the right to say you have leverage.

Now bring this back to the flashlight in fog.

A naive person will test influence like this: blink pattern, see no response, conclude "no influence." A naive person will also test influence like this: blink pattern, see a random shimmer across the field, conclude "influence."

Both are wrong for the same reason: they aren't calibrated. They're reading meaning into a medium that produces shimmering on its own.

Fog scintillates. Air is never perfectly still. Moisture catches light and flickers even when you do nothing. The background produces false positives.

So the only honest test is comparative:

What does the far field look like when you blink?

What does it look like when you don't blink?

What does it look like when you blink patterns that are deliberately detached from any attempt to communicate—random patterns, shuffled patterns, patterns out of alignment with the moments you'd "naturally" choose?

If the far field looks the same across these conditions, then your blinking is not distinguishable from the fog's own shimmer.

That is the boundary: the first distance at which your actions vanish into the baseline variability of the medium.

Notice what this boundary is not.

It is not the first point where you feel powerless. Feelings are noisy. People can feel powerless inside a tight loop where they actually have leverage, and people can feel powerful while pushing on air.

It is not the first point where outcomes become bad. Outcomes can remain acceptable even after leverage becomes hard to detect; the world can coast on inertia for a while. You can still "not crash" even when you're no longer demonstrably steering, because stability can persist without active control.

The boundary is narrower and colder:

It is the first point where your attempted influence can no longer be distinguished from the null baseline you get when the link between action and later outcome is broken on purpose.

At that point, the honest sentence becomes available:

"Under these conditions, at this delay, what I do is statistically indistinguishable from what I would get if I did nothing."

That sentence is not despair. It is a coordinate.

And coordinates are useful because they let you map a landscape.

Here's what the landscape looks like when you stop lying:

At short distances—short delays—your actions can carve detectable differences into what happens next. The medium is clear enough. Your signal survives. The receiver can read your pattern.

As distance increases—delay increases—the medium starts to smear your structure. Fine differences wash out. The world injects variability faster than your signal can dominate it. Your influence score drops toward the baseline shimmer.

And then, at some point, you cross. You can still blink. But the far field stops showing evidence that your pattern exists.

This is the crossing you're going to locate.

And you're going to do it with brutality, because without brutality you will baptize background structure as agency and call it a discovery.

You'll do it by building baselines that preserve the world's real texture. You'll do it by refusing to be impressed by nonzero numbers. You'll do it by setting a threshold that says: "I only claim influence if it exceeds what the null world produces almost all the time."

This is where people get nervous, because it sounds like you're making influence "rarer."

Good. Influence is rarer than the feeling of influence. That's the whole point.

The fog doesn't care how many times you blink. It doesn't care how sincere you are. It doesn't care how correct your prediction of the ridge-walker's path might be. The fog only cares about whether your signal rises above shimmer in a way a receiver could reliably decode.

So the first task is simple to state, hard to accept:

Stop asking "did I act?" and start asking "did the future become detectably different because I acted, compared to a world where my action link is severed but everything else stays real?"

If you can answer that, you can find the boundary.

If you can find the boundary, you can stop mistaking motion for leverage.

And once you stop mistaking motion for leverage, you're ready to name the horizon—not as poetry, but as the first place the world politely stops noticing you.

The first time you try to measure influence honestly, you will feel stupid.

Not because you are unintelligent—because the world is full of cheap structure that impersonates your will. The world is generous with patterns. It repeats itself, drifts, returns, overshoots, settles, and repeats again. It has moods. It has inertia. It has memory. And if you show up with a naive detector—any naive detector—you will "discover" control everywhere.

That is the trap: the universe makes coincidence look like command.

Picture a pendulum in a long hallway. Heavy, slow, dignified. You walk beside it and lightly tap it at the same point in its arc, the way a child taps a swinging door. The pendulum keeps doing what pendulums do: swing. It was going to swing anyway. But your tap is timed to the moment you can predict best—the moment it passes you, the moment it is most visible, the moment your body's rhythm matches it. Your tap correlates with the swing's future position because the swing's future position is already implied by the swing's current state.

So you walk away believing you "guided" it.

Now make the pendulum even more insulting: attach a sensor that reports "influence score" whenever it detects alignment between your tap and the next position of the bob. The score lights up constantly. You are a wizard, according to the meter.

You are not a wizard. You are standing next to a predictable object and touching it at predictable times.

This is what happens in any environment with persistence. Anything with momentum will reward you with false confidence because "the next moment" is largely a continuation of the previous moment. The future resembles the present, so any action aligned with the present will look like it shaped the future.

That's why "nonzero coupling" is meaningless on its own.

Nonzero is easy. Nonzero is the default. Nonzero is what you get simply by existing inside a world that is not white noise.

If you want influence—usable influence, influence that survives delay and noise and cannot be explained away by background drift—you need to compare your claimed power against the world's own habit of producing apparent power when no power exists.

That comparison is calibration.

Calibration is how you stop being impressed by numbers that are merely "not zero."

Because the world will hand you not-zero all day. It will hand you not-zero in traffic. It will hand you not-zero in weather. It will hand you not-zero in crowds. It will hand you not-zero in anything that has continuity, which is almost everything.

So the question you ask is not "is there a signal?"

The question is: **is this signal bigger than what appears when the link between action and later outcome is broken on purpose, while the world's texture remains intact?**

That phrase—"while the world's texture remains intact"—is the spine.

If you break the link by destroying the texture, you are cheating. You are comparing a living ocean to a still photograph and declaring your paddle strong because the photograph never moves.

The ocean must remain an ocean in your baseline.

The tide must still rise and fall.

The waves must still have their lazy memory.

The wind must still push the surface into patterns.

Only the specific thread you claim—your action pulling the future—must be severed.

This is what a "null world" really is: a baseline that keeps the world real and removes only the causal thread you're trying to claim.

A real null world humiliates you because it preserves everything you secretly relied on:

- the tendency of outcomes to resemble their recent past

- the daily rhythm you didn't realize you were timing to
- the long drift you mistook for your gentle steering
- the burstiness in your own behavior
- the slow modes in the environment that make prediction feel like control

A fake null world flatters you because it removes those crutches and leaves you standing beside a featureless void. Against a void, any pattern looks like power. Against a void, your meter lights up and tells you you're special.

So the baseline must be cruel.

It must preserve the world's inertia while cutting the wire that connects your action to later change.

How do you do that without magic? Without cloning the universe?

You do it by rearranging the evidence in ways that keep the texture but break the supposed transmission.

There are two families of rearrangement that matter because they attack different kinds of self-deception.

The first breaks the link by **scrambling actions within comparable moments**.

Comparable moments means: you only swap actions among situations that looked the same to the actor at the time. Same kind of weather. Same kind of local cues. Same kind of accessible record. Same visibility. Same constraints. You don't swap a nighttime decision into a noon scenario. You don't swap an action taken in panic into a calm setting.

You don't compare a tap on a pendulum at its peak to a tap at its midpoint and call it fair.

You keep the "what was visible" context matched.

Then you shuffle the action choices inside that matched context. The world's drift stays. The actor's tendencies stay. The environment remains structured. But any real action→outcome thread is severed because the actions are no longer aligned with the outcomes they originally preceded.

If your influence measure stays high after this shuffle, it is not measuring influence. It is measuring structure that existed without your action mattering.

The second family breaks the link by **shifting time while preserving internal rhythm**.

This is even more humiliating because it keeps your action stream looking exactly like itself: the same bursts, the same fatigue patterns, the same daily cycle, the same twitchy moments where you act three times in a row, the same long stretches where you do nothing. Everything about your behavior remains intact. The only thing that changes is alignment: the action stream is slid forward or backward relative to the outcome stream.

If your influence measure remains high even when the alignment is wrong, then your influence measure is reading shared rhythms—things both streams inherit from the same drifting world—rather than transmission.

This is the purest test against the "I was riding the current" illusion.

Because the current is real. The current will make both your actions and outcomes look patterned. The current is what makes prediction possible. But the current is not your steering.

A time shift keeps the current and destroys the steering thread.

So a serious claim of influence must beat both kinds of baselines: the shuffle that respects comparable moments, and the shift that respects temporal texture.

Once you adopt these baselines, you discover a frightening fact:

A lot of "agency" disappears.

Not because you became weaker. Because you stopped counting counterfeit.

Because you stopped praising yourself for being aligned with a world that was already going that way.

This is where people panic and try to bargain. They say: "But my effect is still there, it's just small."

Small is not a defense. Small is a description.

If your measured influence is the same size as the world's baseline counterfeit under null rearrangements, then you don't get to claim leverage. Not in a usable sense. Because you can't rely on it. You can't tell when it's real and when it's shimmer. You can't build a robust action policy around it without fooling yourself.

So you must set a bar.

Not an arbitrary bar. A calibrated bar.

You look at the distribution of fake influence produced by your null worlds. You see how large your metric gets when there is no transmission—only structure, only drift, only shared rhythms.

Then you choose a tolerance for embarrassment.

You choose how often you are willing to declare influence when none exists.

A cautious person chooses a high bar. They require their measured influence to exceed what appears in almost all null worlds. They accept that sometimes the world will still fool them. They simply refuse to pretend otherwise.

This is where the boundary begins to take shape.

Because once you have a calibrated bar, you can do something that feels like sacrilege:

You can watch influence die as delay increases, and you can say exactly where it dies.

You do not declare influence dead because the number becomes "small." Small compared to what? Your mood?

You declare influence dead because the number falls into the region where the null worlds live. Into the baseline shimmer. Into the fog's own glitter.

At short delays, real influence can stand above the null shimmer—if it exists, and if you aren't just touching the meter.

At longer delays, the medium smears the signal. Noise and drift swallow fine structure. The action thread becomes harder to trace. The measured coupling sinks.

Eventually it sinks below the calibrated bar. Not once, not in a single twitchy crossing that might be estimator wobble, but persistently, across a range of delays.

That is the boundary in operational clothing:

The first distance where your action becomes indistinguishable from null influence, given the record you actually had at the moment of choice.

Notice what this does psychologically.

It turns "powerlessness" into a coordinate. It turns despair into a map.

It also forces you to face a distinction that will matter later:

You can still be predictive in the fog.

The ridge-walker still has a path. The river still has a current. The pendulum still swings. The world remains structured, so forecasts remain possible. Your eye for drift can remain sharp even while your transmission falls below the bar.

So you can keep being right.

But now you no longer confuse being right with being able to change.

Because you have a calibrated baseline that shows you how easy it is to look powerful in a structured world even when your signal never crosses the medium.

The fog doesn't care about your sincerity. It cares about distinguishability.

And once you accept distinguishability as the criterion, you're ready for the next move: turning this into an actual horizon, not as poetry, but as the first delay at which your measured influence sinks into the null floor and stays there.

A foggy field will teach you something that spreadsheets can't: the world can produce shimmer that looks like signal even when you're doing nothing at all.

If you stand still long enough, you'll see it. Moisture drifts. Air swirls. A distant lamp trembles through vapor. Your eyes invent structure. Your brain announces meaning. You didn't send anything, yet the field still flickers as if it's answering. That's the baseline. That's what "nothing" looks like in a living medium.

Now imagine you blink your flashlight again. You want to know whether the far field is responding to you or merely continuing its own shimmer. This is the real test of influence: not whether the world moves, but whether the world moves *differently* because of you.

To answer that, you need a bar. A line. A threshold that says: "Above this, I will claim my signal survived. Below this, I will assume the medium is just being itself."

People hate thresholds because thresholds force accountability. They force you to admit you can be fooled. They force you to pick how often you're willing to be wrong in a specific direction: declaring influence when there is none.

But without a threshold, you can always tell yourself the faintest twitch means you still matter. Without a threshold, you can hold onto a dying signal forever, cradling it like a superstition.

So you build a threshold from null worlds.

You take your most honest baselines—ones that keep the world's drift intact while severing the action thread—and you measure how much "fake influence" they produce. You do it many times. You let the medium show you how often it can accidentally look responsive.

The result is not a single number. It's a spread. A whole landscape of counterfeit coupling. Some null runs produce almost nothing. Some produce a surprisingly bright shimmer. Most produce something in between.

That spread is the medium's own capacity for lying to you.

Now comes the key move: you decide where you draw the line.

Not by vibe. By tolerance.

You choose how rarely you want to be fooled by baseline shimmer. Maybe you say: "I will only claim influence if what I measure is higher than what I see in almost all null worlds." That means you're demanding that your signal be unusually strong relative to the medium's habitual flicker. You're making it expensive to claim power.

This is the first place the whole framework turns from storytelling into instrumentation.

Because now you can say something that doesn't rely on your ego's need to feel effective:

"I will call this 'leverage' only if it exceeds the level the world routinely produces when leverage is broken on purpose."

That's it. That's the bar.

And once you have that bar, the horizon becomes visible.

You take the delay between action and outcome and you walk it forward—step by step—while applying the same calibration at each step. You keep the accessible context fixed. You keep the baseline procedures honest. You keep the bar consistent. And you watch what happens to your measured influence as you demand it survive longer transport through the medium.

At short delays, you might see a clear signal above the bar. At longer delays, the signal usually shrinks, because the medium has more time to smear it, because noise has more time to swamp it, because the world has more opportunities to do its own thing and bury your contribution.

Eventually, you reach a region where your measured influence lands inside the null spread. Not once, not as a fluke, but as a habit. And that's the moment you earn the right to say the sentence that most people spend their lives avoiding:

"Beyond this delay, what I do is indistinguishable from baseline."

That is the horizon. Not poetry. Not a spiritual boundary. A calibrated crossing point.

Now, there are three ways people try to cheat this moment, and you have to cut each one off before it becomes a religion.

First cheat: "But it's still nonzero."

Nonzero means nothing in a structured world. A structured world produces nonzero all by itself. The fog produces shimmer. The river produces drift. A market produces autocorrelation. A crowd produces waves. If your metric returns nonzero under null conditions, then nonzero is the default. It's not evidence of transmission. The bar is not zero; the bar is the null world's typical shimmer.

Second cheat: "It crossed the bar once."

One crossing is not a boundary. One crossing is often estimator wobble, a lucky alignment, or a rare null-like event that accidentally looks significant. If you allow single-point crossings to define the boundary, you will spend your life chasing mirages. You'll declare "agency is back!" every time the fog thins for a moment and then feel betrayed when it thickens again.

So you need persistence.

You require the signal to stay below the bar in a sustained way before you declare the horizon reached. Not because you're moving goalposts, but because the medium is noisy and measurement is finite. A real boundary is not a single dramatic step; it's a region where the signal no longer reliably clears the bar.

Persistence is the difference between science and gambling. It says: "I'm not going to call it dead because it dipped once, and I'm not going to call it alive because it spiked once."

Third cheat: "I don't like this null."

Of course you don't. A good null hurts. A good null is designed to preserve the very structures you relied on unconsciously. If your null destroys structure, you'll look powerful by comparison. That's the flattering null. The honest null preserves drift, burstiness, inertia, daily rhythm—whatever texture your world has—and severs only the action thread. If you don't like it, that's usually because it's doing its job.

The horizon is not defined by what you wish to be true. It's defined by what survives calibrated baselines.

Now let's make this visceral with a scenario that doesn't require any equations.

Imagine a remote crane operator again, but now you're not watching the hook in real time. You're watching a delayed video feed. At short delay, you can still guide the hook with care. Your corrections land soon enough that the hook responds in a way that's readable. Your influence is detectable. It clears the bar because when you randomize your actions under matched conditions, the outcomes stop aligning the way they do when you truly steer.

Now increase the delay. The hook becomes a prophet of your past, not a partner of your present. You correct what you saw a second ago, but the hook has already moved. You correct again, and now you're overcorrecting. Your actions start to correlate with large swings—not because you're steering, but because you're injecting oscillation. In that regime, the world still moves a lot. It may even move more. But your intended influence—your ability to place the hook precisely—begins to vanish.

Here's where naive measurements will lie.

If you measure "effect" as raw movement, you'll conclude you have massive influence, because the hook is moving dramatically in response to your frantic inputs. That's the meter-twisting version of agency: you can make the display dance by thrashing.

But if you measure influence as "did your actions shift the outcome distribution in the desired way relative to matched baselines," you'll see the truth: your inputs no longer reliably shape the future state you care about. They blend into the turbulence. Under null rearrangements that preserve the hook's inertia and your bursty behavior, you'll see similar chaos. Your "control" has become indistinguishable from the medium's own instability.

That indistinguishability—calibrated and persistent—is the horizon.

Now notice something subtle and important: the horizon is not the moment the crane starts failing catastrophically. Sometimes you can keep the hook "good enough" for a while even after your measurable influence has dropped below the bar, because the world can coast. Inertia can carry you. The environment can be forgiving. A task can tolerate sloppy placement for a stretch.

This is a crucial distinction: losing detectable influence is not always the same as immediate failure. You can be "functioning" while already beyond the boundary of demonstrable leverage. You can be surviving on the world's momentum.

That is why this boundary is so psychologically difficult. It doesn't announce itself with a crash. It announces itself with a quiet statistical statement: you can no longer prove your actions matter beyond what baseline structure would produce anyway.

People call that depressing. It isn't. It's clarifying.

Because once you can locate the horizon, you can map how it moves.

You can push it outward by improving the medium: reducing noise, tightening feedback, increasing the meaningfulness of each action, cleaning the record you condition on. You can also watch it collapse inward when the environment becomes turbulent, feedback becomes stale, or measurement becomes contaminated by readout tricks.

This is the turn where "power" stops being a moral drama and becomes a geometry of constraints.

The horizon is the place where your flashlight beam becomes fog bloom. The place where your careful blink pattern becomes shimmer. The place where your action becomes observationally equivalent to silence—not because you are weak, but because the medium cannot carry the fine structure of what you're trying to transmit.

And once you accept that, you're ready for the next move: making the boundary robust against noise by requiring persistence, and making it honest by choosing null families that preserve the world's texture without preserving your fantasy.

That's where the horizon stops being a metaphor and becomes a coordinate you can defend under hostile questioning.

Noise doesn't let you have clean endings.

That's what makes it such a perfect accomplice for self-deception: it gives you little miracles—tiny spikes, brief clearings, sudden "proof" that you still have leverage—right when you're most desperate to believe.

In the fog, there are moments when the air thins. The beam carries a little farther. A distant shape sharpens. Your blink pattern feels readable again. You feel that dangerous warmth in your chest: *it's back.*

Then the fog thickens, and the warmth turns into resentment: *it was taken from me.*

But the medium didn't "give" and "take" anything. The medium is jittery. Your sampling is finite. The world is structured, and structure produces rare alignments that impersonate transmission. That's why a boundary defined by a single crossing—one moment where your measured coupling falls below your null bar, or one moment where it rises above—will turn into superstition. You'll spend your life calling miracles and betrayals from the same underlying randomness.

So the next step is to turn the boundary into something that can survive the medium's flicker.

The boundary has to be robust.

Robust does not mean "complicated." It means "not fragile to one lucky coincidence."

The most common fragility looks like this: you evaluate influence at many distances (or many delays). You do it honestly. You keep your baselines structured. You enforce time separation. You condition on what was visible. You even set a high bar drawn from null worlds.

And still—because you are checking many distances—one of them will occasionally cross the bar by luck.

That's not a failure of the method. That's a fact about searching.

If you stare at enough slices of a noisy medium, one slice will look like a message. The fog will glitter in just the right way. The river will eddy in just the right way. The drift will line up with your action stream in just the right way. If you allow single-slice miracles to define leverage, you will always find leverage, even where none exists.

The cure is persistence.

Persistence is how you stop treating lucky alignments as evidence.

Instead of asking, "Did my influence clear the bar at this one distance?", you ask, "Does it clear the bar reliably across a stretch of distances?"

And instead of asking, "Did my influence fall below the bar once?", you ask, "Did it sink into the null floor and stay there?"

The boundary becomes not a single pixel but a band: a region where the medium consistently fails to carry your signal above baseline shimmer.

This is exactly how good detectors behave in the physical world.

A smoke alarm doesn't go off because a single dust mote crosses its sensor. A heart monitor doesn't declare catastrophe because one heartbeat is noisy. A radar system doesn't launch a response because one blip appears in one frame. In all those cases, a single-frame spike is expected under noise, and overreacting to it makes the device useless.

So detectors demand repeated evidence: multiple consecutive frames, or a majority of frames in a short window, or a sustained elevation above background.

That's persistence.

It's not "moving the goalposts." It's acknowledging that the medium flickers.

Now translate that into the language of influence.

You're looking for the first distance where your measured coupling becomes indistinguishable from null. But "indistinguishable" is itself a noisy judgement in finite data, because your coupling estimate has variance. Your null threshold has variance. The world is nonstationary.

The record pipeline wobbles. Even under perfect honesty, the line will jitter.

So you don't declare the boundary at the first jitter. You declare it at the first sustained collapse.

A simple persistence rule looks like a human sentence:

"Starting here, for a while, the measured influence fails to beat the null bar most of the time."

That's a robust boundary.

And there are different ways to enforce it, each with a slightly different personality.

One way is a run-length rule: you require a streak. For example, you don't declare the boundary reached until you see several consecutive distances where influence fails to clear the bar. The streak length should be long enough that it's unlikely to occur by chance under genuine influence, but not so long that it becomes impossible to declare anything in noisy real-world data.

Another way is a majority rule: you define a short window of distances and require that most of them fall below the bar. This has a useful property: it tolerates occasional spikes—those little miracles that the fog produces—without letting them dominate the conclusion. A single spike does not resurrect leverage if the surrounding region is dead.

A third way is a two-stage rule: you first identify a candidate boundary where the signal starts failing, then you confirm it with an independent slice of data or an independent time segment. This is the harshest, because it punishes overfitting—finding "agency" that exists only in one sample path. But it's also the most convincing under hostile scrutiny, because it replaces "I saw it once" with "I see it again."

All of these methods share the same ethic: you refuse to let one noisy coincidence define your reach.

This matters more than it seems, because the human mind is built to worship spikes.

A spike feels like revelation. A spike is the story you tell yourself: "I still have it." And once you've told that story, you will begin to tune your behavior around it. You will act as if leverage exists at that distance. You will design strategies for it. You will blame yourself when it fails again.

But if the spike was merely baseline shimmer, then your strategy is built on fog glitter. You will suffer for a leverage you never had.

Persistence protects you from that suffering by refusing to grant the spike moral authority.

Now, persistence alone isn't enough if your baseline is weak.

If your null worlds don't preserve the world's real texture, you'll set the bar too low and everything will "persist" above it. You'll manufacture fake robustness. You'll build a stable illusion.

That's why the null family matters just as much as the persistence rule.

The structured shuffle-type baseline attacks the "shared conditions" illusion: it keeps comparable moments comparable, then scrambles the action thread so any real transmission should vanish.

The structured shift-type baseline attacks the "shared rhythm" illusion: it preserves the temporal texture of both streams while breaking alignment, so any real transmission should vanish.

If your claimed influence clears both bars, that's harder to counterfeit. If it clears only one, you've learned something valuable: you may be measuring a particular kind of structure rather than transmission.

In a living medium, you should treat null families like different interrogation methods. Each one forces the truth to survive a different kind of skepticism.

Now comes another subtle trap that persistence helps with: selective attention.

Suppose you evaluate many distances and you see a few that clear the bar. You feel tempted to focus on those. To declare, "This is where my influence lives." To build a story about resonance—why your action reaches exactly this far and not farther.

Sometimes that story is real. Sometimes the medium has a regime where transmission is better. Sometimes the channel does have a ridge where influence peaks before collapsing.

But sometimes you are merely looking at the natural peaks of noise. And if you don't enforce persistence and cross-check against multiple null families, you can build a cathedral on a coincidence.

So you need a discipline of humility:

- You treat isolated clearings as suspect.
- You treat sustained regions as evidence.
- You treat anything that appears only under one baseline as incomplete.

- You treat anything that disappears when you shift the time alignment as shared rhythm.
- You treat anything that disappears when you shuffle within comparable contexts as shared conditions.
- You treat anything that survives all of that as the best candidate for real leverage.

This is not paranoia. It's the price of making a claim about influence in a world that loves to produce illusions.

Now, once you enforce persistence, the boundary becomes easier to see—and it starts to look like something you can map.

At shorter distances, real leverage—if it exists—tends to produce coupling that clears the calibrated bar consistently. You get a region of reliable distinguishability.

As distance increases, the coupling begins to flicker. It clears the bar sometimes, fails other times. This is the transitional band—the fogline. It's not yet a horizon, but it's where the medium is beginning to win.

Then, past a certain point, the coupling falls into the null floor and stays there most of the time. Occasional spikes may still appear—fog glitter, lucky alignments—but they don't persist. They don't survive the window. They don't survive the second baseline. They don't replicate.

That sustained collapse is the boundary you can defend.

It's also where the emotional language starts to change, if you let it.

You stop saying "I lost control," which sounds like a moral failure.

You start saying, "Beyond this distance, I cannot demonstrate that my action changes the downstream distribution beyond what baseline structure produces anyway."

That sentence sounds cold. It should. Cold is what keeps you honest.

And now a final twist that makes this boundary even more important:

Sometimes performance remains acceptable beyond the boundary.

The boat keeps floating. The crane doesn't crash. The temperature stays "fine." The outcome remains within tolerances. That can fool you into thinking you still have leverage.

But tolerable outcomes can be carried by inertia. The medium can coast. The environment can be forgiving. A process can remain stable without your steering.

So you need the boundary not because everything fails immediately when you cross it, but because crossing it means you are no longer distinguishably steering—whatever success remains is riding drift. This is the cruelest kind of quiet failure: you can look competent while your influence is already below the floor.

Which is why a robust horizon must be defined by distinguishability, not by catastrophe.

A catastrophe is loud and late. A boundary is quiet and early.

So the boundary you're locating here is not the moment everything breaks. It's the moment your signal stops being readable above shimmer in a way that persists.

And once you've done that—once you've defined the boundary in a way that doesn't flinch at spikes—you can do something that feels like power again, in the only honest sense:

You can move it.

Because now you can see what shrinks it and what extends it: noise, delay, record quality, action bandwidth, feedback tightness. The boundary stops being fate and becomes a coordinate with knobs you can measure.

Noise doesn't give you clean borders. It gives you flicker.

That's the entire problem, and it's also the entire reason people get trapped in false hope. A medium with variability will occasionally produce a moment that looks like a message. A delayed world will occasionally line up in a way that makes your last action appear prophetic. A structured environment will occasionally reward you with a perfect coincidence. If you build your definition of "reach" around those moments, you will live inside a casino of your own interpretation.

You will mistake glitter for signal.

So the next move is not a new concept. It's a discipline: **don't let a single lucky moment define reality.**

Think about radar on a rough sea. The screen shows a field of speckle—tiny fluctuations from waves, spray, weather, reflections. Sometimes the speckle forms a shape.

Sometimes it forms a streak. Sometimes it forms something that looks exactly like a boat moving fast.

If you declare "target acquired" every time the speckle makes a boat shape once, you will be wrong constantly. Not because the sea is evil, but because the sea has texture. Texture generates false targets. The more frames you watch, the more false targets you'll see.

A competent operator doesn't worship single frames.

A competent operator demands persistence: the "target" must appear again, and again, and again, in a way that is coherent under time. It must behave like a thing, not like a coincidence.

That's how you treat influence once you stop flattering yourself: the evidence must persist.

Because the moment you start checking many distances (or many delays), you create a new trap: **the more chances you give noise to fool you, the more it will.**

This is basic and brutal. If you look in one place, you might see nothing. If you look in a hundred places, you will almost certainly see *something*. That "something" will tempt you into a story.

"I found it. This is where leverage lives."

Sometimes that story is true. Often it's a selection effect: you didn't find a signal; you found the maximum of noise.

So your boundary cannot be "the first time I fail." It also cannot be "the last time I succeed." Both are too sensitive to flicker. Both can be hijacked by single-point accidents. You need a boundary definition that treats the medium like a medium: variable, structured, and eager to impersonate transmission.

That's what persistence does. It replaces single-point drama with sustained behavior.

Instead of asking, "Did my measured coupling clear the bar at this distance?" you ask, "Does it clear the bar **reliably** across a stretch?" And instead of asking, "Did it fall below once?" you ask, "Has it collapsed into the null floor **for long enough** that brief spikes are clearly just glitter?"

This is not pedantry. This is the difference between seeing a ship and seeing a wave.

Here are the three common ways flicker ruins honest measurement, and the corresponding persistence rules that repair it.

The first ruin: the "miracle spike"

You measure influence across delays. Most points look dead. One point shoots up above your null bar. It looks clean. It looks like proof. Your brain lights up: *It's back.*

But if that spike does not replicate in nearby delays, it is almost always a coincidence: an accidental alignment between your action stream and an outcome stream that already had structure. The medium produced a bright shimmer.

A persistence rule treats the spike as what it is: interesting but not sufficient.

The simplest form is a streak requirement: you don't accept leverage at a distance unless you see multiple adjacent distances clear the bar, not just one. Why adjacent? Because real transmission doesn't usually appear at a single isolated delay and vanish immediately; it tends to occupy a region—thin or wide—depending on the medium. Isolated spikes are a signature of noise.

A more forgiving version is a majority-in-window rule: within a small window of neighboring delays, you require that most points clear the bar. This tolerates occasional dips (estimator wobble) while still rejecting lone miracles.

The second ruin: the "false funeral"

The inverse also happens. You're watching your influence curve, and at some delay it dips below the bar once. You panic. You declare the boundary reached. You build a story of loss.

Then the next delay clears the bar again. And the next. And now you're whiplashed: dead, alive, dead, alive. You start treating the medium's normal jitter like a moral narrative.

A persistence rule prevents premature funerals. You do not declare the boundary reached on the first dip. You declare it reached when the curve stays below the bar for a sustained region—again using a streak rule or a majority-in-window rule.

This is the same logic a heart monitor uses. A single noisy beat does not define collapse. Sustained deviation does.

The third ruin: the "multiple-slices curse"

Even with good null baselines, if you evaluate many delays, some will look "significant" by chance. This is not a flaw; it's mathematics of searching. The more slices you inspect, the more likely you are to find an extreme value. If you pick that extreme value and tell a story about it, you are writing fiction that looks like analysis.

Persistence helps, but you also need a habit: **don't let your detector's most exciting discovery be the only evidence.**

A strict approach is to demand replication in an independent segment. Not because life is a laboratory, but because selection effects are real. If you "discover" leverage at a certain delay in one run, you confirm by checking whether it appears again in a different run, or a different time block, or a different seed, or a different slice of the environment. If it vanishes, it was likely glitter.

This is the closest you can get to cloning the universe: you demand that a claim survives being asked twice.

Now, there's a subtle beauty here: persistence rules do not add complexity for its own sake. They encode a very human truth:

A real channel effect is coherent. Noise is not.

Coherence is what you're searching for. Coherence across time, across neighboring delays, across independent blocks, across null families.

And that last phrase matters: across null families.

Because you can build a baseline that is too gentle without realizing it. You can choose a null that breaks the link in a way that accidentally leaves a loophole for shared rhythms to impersonate transmission. You can choose a null that preserves one kind of structure but not another, and your detector can be fooled in a consistent way.

So the strongest posture is to demand that your influence survives more than one kind of skepticism.

One null family attacks "shared conditions" illusions: your action is correlated with the outcome because both respond to the same visible context. A baseline that shuffles actions among comparable contexts breaks that.

Another null family attacks "shared rhythm" illusions: your action stream and the outcome stream both inherit slow drift and periodicity, so they align even without transmission. A baseline that time-shifts while preserving internal texture breaks that.

If your measured influence clears the bar against both families—persistently—then you have something that is much harder to counterfeit. If it clears one but not the other, you haven't failed; you've learned which illusion you were living in.

Now look at what persistence and multi-null skepticism do to the boundary itself.

The boundary stops being a dramatic cliff. It becomes a **transition band**:

- A region where influence is reliably distinguishable from baseline shimmer.
- A region where it flickers—sometimes distinguishable, sometimes not.
- A region where it sinks into the null floor and stays there, with occasional glitter spikes that do not persist.

This transition band is exactly what a real medium produces. The fog line is not a knife-edge. It's a thickening.

The sea clutter doesn't stop; it dominates. The signal doesn't flip off; it becomes unreadable.

So the boundary you define is not "the first moment things look bad." It's "the first point after which the medium behaves as if you're silent, most of the time, under calibrated skepticism."

That's a mouthful, but it's the only honest definition that doesn't crumble under pressure.

And it has a payoff that most people miss because they're addicted to drama:

A boundary defined this way can exist **before** visible catastrophe.

You can lose demonstrable influence while outcomes remain acceptable because inertia carries you. A plane can remain stable with a bad controller for a while because the airframe has damping. A market can remain calm while leverage silently dies because regime inertia persists. A relationship can "look fine" while meaningful influence has already dropped below the floor because the routine keeps running.

That's why this boundary matters. It's early. It's quiet. It catches the loss of reach before the crash makes everyone honest.

It also gives you a real knob to turn: you can ask what moves the boundary.

If you tighten feedback, does the region of distinguishable influence extend? If noise increases, does it collapse inward? If your action bandwidth improves—not in frantic frequency, but in meaningful impact per intervention—does coherence return? If your record quality improves, does conditioning remove more counterfeit and reveal a cleaner region where transmission truly exists?

This is the moment "power" stops being an identity and becomes an experiment.

You are no longer asking, "Am I powerful?" You are asking, "Under what conditions does my signal remain readable above shimmer, and for how long does it persist before the medium turns it into fog bloom?"

That question is cold enough to be useful. It refuses single-frame miracles. It refuses single-frame funerals. It refuses stories built on the maximum of noise.

It treats the world like a medium with texture and your influence like a signal with a finite reach. Which means you're finally ready for the next step: turning that transition band into a single defendable coordinate—by choosing the persistence rule, choosing the tolerance, and stating the boundary as the earliest delay where sustained null-consistency begins.

There's a moment where all the metaphors stop helping and you have to pick a number.

Not because reality is made of numbers, but because **you** are made of constraints. Finite attention. Finite time. Finite patience. Finite tolerance for being fooled. If you don't choose a threshold, you will keep reinterpreting shimmer as signal forever. You'll keep "seeing" leverage in every little bright accident the medium hands you, and you'll call those accidents evidence because you want them to be evidence.

A threshold is not an act of arrogance. It's an act of surrender to the fact that the medium flickers even when you do nothing.

Stand in the fog with your flashlight turned off. Watch long enough and you'll still see "events." A distant lamp flares slightly as moisture passes. Your own eyes micro-jitter. Your brain clusters randomness into shapes. Even silence produces a kind of chatter.

That chatter is your baseline. That's the world's free counterfeit.

So when you turn your flashlight on and blink a pattern, the real question isn't "did anything happen?" Something always happens. The question is:

Did what happened exceed the kind of 'something' the medium produces by itself?

This is where the null worlds stop being a philosophical tool and become a measuring stick. You take your honest baselines—the ones that keep the world's texture intact while cutting the action thread—and you look at the distribution of counterfeit "influence" they produce.

It will not be zero.

It will not even be small in the way your ego wants small to be. It will have a fat tail. Sometimes the medium will produce a startlingly bright shimmer that looks like a response. That's why single spikes are poisonous. That's why persistence matters. And that's why your threshold has to be built from the null distribution rather than from your hope.

Now: how do you set the bar?

You don't set it by asking, "What feels significant?" You set it by asking a more honest, uglier question:

How often am I willing to be fooled into declaring leverage when the causal thread is severed?

That is what a false alarm rate is. It's not abstract. It's a moral contract with your future self.

If you set the bar too low, you'll live in a world where your detector is constantly shouting "influence!" even in null worlds. You'll be a prophet of your own mirages. Everything will look steerable. You will feel powerful. You will also crash into reality repeatedly and blame yourself for not steering well enough, when the actual failure was that you declared steering rights you never had.

If you set the bar too high, you'll live in a world where you refuse to claim leverage even when it exists, because the medium won't let you demonstrate it under your strict standard. You'll become cautious to the point of paralysis. You'll treat real, modest influence as if it were fog glitter. You'll miss opportunities because you demanded proof stronger than the situation can provide.

So the bar is a trade. Always.

The cleanest way to make the trade explicit is to choose a percentile of the null distribution—spoken in plain language, not symbols.

For example:

- "I will only claim influence at a given delay if my measured coupling is higher than what the null world produces **99 times out of 100**."

That sentence means: if you take the null world—where you've severed the action thread—and you run it many times, only about 1% of those runs will produce a counterfeit influence measure as large as what you observed. Your observed value is rare under the baseline shimmer.

This is how you earn the right to say: "This looks like real transmission."

It's not proof in a cosmic sense. It's a refusal to be easily fooled.

Now here's where people usually mess it up: they pick one threshold and treat it like a universal constant.

But the medium's shimmer can change with delay.

At short delays, there are more opportunities for contamination: readout-layer shortcuts, overlapping windows, logging artifacts. Even with careful hygiene, the null baseline can be different there because the world and your pipeline behave differently at different time scales. At long delays, the world's inertia and drift can create slow, persistent structure that generates counterfeit coupling in different ways. The baseline shimmer isn't flat; it has shape.

So if you want to be honest, you set the bar **as a function of delay**. Not because you're being fancy, but because you're admitting the medium is not uniform across distance.

This is also why comparing your observed coupling to "zero" is childish. Zero is not the baseline in a structured world. Zero is a bedtime story you tell yourself so you can claim influence without doing the hard work.

The baseline is what your null worlds produce.

The bar is what they produce **rarely**.

Now add the next insult: scanning many delays increases the chance of finding a bright accident.

If you look at one delay, the odds that your measured coupling beats the null bar by luck might be small. If you look at fifty delays, the odds that at least one of them beats the bar by luck rises dramatically. This is not a flaw in your detector; it's the mathematics of giving noise more lottery tickets.

So you need a second contract with yourself:

I will not let a single rare event define leverage when I have searched widely.

That's where persistence enters again, not as decoration, but as protection against "found it!" syndrome.

You don't claim influence at a delay because it cleared the bar once. You require it to clear the bar repeatedly across nearby delays or across nearby windows, or to replicate in a second segment. Likewise, you don't declare influence dead because it dipped once; you require sustained collapse into the null band.

This combination—percentile bar plus persistence—does something psychologically important: it makes your detector boring.

Boring is good. Boring means you're not gambling.

A detector that is easily excited will make you feel alive. It will also make you wrong constantly. A detector that is boring will make you slightly uncomfortable because you can't feed on constant "signals." It will also keep you sane.

Now, once you set the bar, you can define the boundary in a way that stops being rhetorical.

You walk forward in delay. For each delay, you compute your influence measure—whatever your chosen coupling statistic is—conditioned on what was accessible at the moment of choice. Then you compare that value to the null bar at that delay.

At short delays, you may see consistent bar-clearing. That's a region of demonstrable leverage. It means your action thread is strong enough to show up above baseline shimmer.

Then you hit the transitional band: bar-clearing becomes intermittent. The medium is starting to win. Your influence is not reliably readable.

Finally you reach the region where your measured influence lives inside the null band most of the time. It no longer clears the bar except for occasional glitter spikes that don't persist. That's where the medium behaves as if you're silent, at least as far as your chosen outcome variable is concerned.

The earliest point at which this "null-consistent" behavior begins—under your persistence rule—is your boundary.

And notice what it depends on:

- It depends on your chosen outcome. You might lose demonstrable influence for one outcome while still retaining influence for another. A controller might lose the ability to place a hook precisely while still being able to prevent catastrophic collisions. Different outcomes have different sensitivity to action.

- It depends on your null family. If your baseline does not preserve the world's relevant texture, your bar will be wrong. That's why you use more than one null family: different baselines attack different counterfeit sources.
- It depends on your tolerance choice. A stricter bar produces an earlier boundary; a looser bar pushes it outward. There is no metaphysically correct choice, only choices that are honest about the cost of false confidence.
- It depends on sampling. With more data, you can detect smaller influences above the null shimmer. With less data, genuine influence can drown inside the baseline spread. The boundary you can demonstrate is not always the boundary that exists in principle; it's the boundary you can see with finite evidence.

This last point is important, because it prevents the boundary from turning into doom.

The boundary is not "the universe forbids influence beyond here." The boundary is: "given this noise, this delay, this action bandwidth, this record quality, and this evidentiary budget, I cannot distinguish my influence from baseline beyond here."

That's a statement about demonstrability under constraints. It's a coordinate.

And coordinates can move.

But you don't move them by shouting. You move them by changing the medium: cleaner feedback, less delay, stronger interventions per action, better conditioning so you stop fighting ghosts created by your own record pipeline.

There is also a darker use of the threshold concept that you should name explicitly, because it happens everywhere: you can control someone by controlling their bar.

If you flood their environment with noise, you raise the baseline shimmer. Their actions become harder to distinguish. If you delay their feedback, you smear their signal. If you degrade the quality of what they can observe, they can't condition properly, so shared-state drift impersonates causality and they misattribute outcomes. Under those conditions, even if they "act," they can't prove their action mattered. Their confidence collapses. They become easier to steer precisely because their own steering cannot clear the bar.

This is not mysticism. It's what happens when you change the medium.

So choosing a null-calibrated threshold is not just a technical flourish. It's the line between:

- living inside the medium's shimmer and calling it leverage, and
- demanding that leverage clear a standard the medium itself rarely fakes.

That line is the beginning of a real boundary.

And once you have a boundary you can defend, you are ready for the next move: distinguishing "loss of demonstrable influence" from "functional failure," and showing how a process can keep coasting on inertia even after leverage is statistically gone—until the cliff arrives.

There is a particular kind of terror that arrives only after you get serious about thresholds.

Not the terror of chaos. Chaos is honest. Chaos at least tells you it's chaotic.

The terror is **quiet competence**.

You can keep doing the thing. You can keep "operating." You can keep hitting targets that are big enough and slow enough. You can keep getting away with it. You can even look smooth.

And all the while, the actual coupling between what you do and what happens later can already be gone.

This is the part nobody wants to hear because it destroys the oldest comfort fantasy: *If I were truly powerless, I would immediately fail.*

No. Often you don't fail. Often you coast.

A heavy object rolling on a flat surface doesn't need steering to keep moving in the same direction. It needs steering only when the world asks it to turn.

That's the first separation you have to let into your bones:

There is a difference between **losing demonstrable influence** and **visible breakdown**.

They can happen at different times. Sometimes very different.

Picture a car on a straight highway at night. The steering linkage breaks—quietly, cleanly. The wheel still turns in your hands. The dashboard still lights up. The tires still hum. The car keeps going straight because the road is straight and the vehicle has momentum and the lane is forgiving.

You can drive like this for miles without realizing anything is wrong, because the world isn't currently demanding a turn. You might even congratulate yourself for your steady hands. You might attribute the smoothness to your skill.

Then a curve arrives.

You turn the wheel. The wheel turns back. The car keeps going straight.

This is when people finally say the words they avoided: *I'm not connected.*

But notice what happened: the actual loss of steering occurred earlier. The *discovery* happened later, when the environment demanded proof.

That's exactly how the boundary behaves in many real settings. It is often crossed quietly, while outcomes remain acceptable due to inertia, damping, redundancy, and the sheer laziness of the world. The cliff arrives when conditions change fast enough that coasting is no longer sufficient.

So when you map influence as "distinguishable from null," you are often locating the moment your steering becomes equivalent to silence **even if the task has not yet collapsed.**

This is not a paradox. It is one of the most important properties of embedded life: *stability can persist without control.*

A ball in a shallow bowl will return toward the center even if you stop correcting it. Not because you're steering, but because the geometry is forgiving. Many processes have this kind of passive stability. They "want" to return to a basin of attraction. A small disturbance dies out on its own. The world does some of your work for free.

That free work is the reason coasting feels like agency.

Now add the second insult: delay.

Delay means your corrections land late. Late corrections often don't cause immediate disaster. Sometimes the environment is slow enough that your late correction still helps. Sometimes the environment is stable enough that your late correction doesn't matter. In both cases, you can remain "functional" while your actions are no longer measurably shaping the downstream distribution in a way that clears calibrated baselines. You can look competent while the coupling is already below the floor.

This is why a boundary defined by catastrophe is useless. Catastrophe is late. Catastrophe is loud. Catastrophe arrives after the opportunity to understand the medium has already passed.

A boundary defined by distinguishability is early. Quiet. Often invisible to anyone who insists on drama.

So you need two kinds of limit in your head—two different endings that people keep confusing:

1. **The detectability boundary**: the point beyond which your action's downstream effect is not distinguishable from calibrated baselines. Your signal is no longer readable above shimmer.
2. **The functional breakdown**: the point where the task fails visibly by whatever criterion you care

> about: collision, runaway oscillation, missed target, unacceptable error rate, cascading instability.

These two are cousins, not twins.

Sometimes they align. Sometimes you lose demonstrable influence right as you fail. That happens when the environment is unforgiving, when inertia is low, when stability is weak, when the medium is turbulent. A curve arrives immediately. The cliff is right after the line.

But very often, they separate. There is a gap where you are already beyond demonstrable steering, yet the task is still "working," because it is being carried by drift, by passive stability, by momentum, by regime persistence, by sheer luck.

That gap is the most dangerous place for the human mind.

Because it encourages a lie that feels reasonable: *If it's still working, I must still have control.*

No. It can be working because the world is temporarily easy.

The honest test is not "is it working?" The honest test is "does my action change what happens later, beyond what the null world produces, under the same accessible context?"

That test can fail while "working" continues. That's the quiet competence terror.

And now we reach the strange situation that defines a lot of modern experience:

You can remain highly predictive while your steering is gone.

The road ahead is visible. The curve is visible. You can forecast the crash. You can even timestamp it. Your prediction channel can stay strong because prediction rides on structure: inertia, drift, slow modes, regimes.

Meanwhile your influence channel is gone because influence requires the world to carry your signal forward without smearing it into noise. Influence requires bandwidth and low latency and an environment that doesn't drown you.

So you can be trapped in a state of cruel clarity: the future is legible, but not pliable.

That state is what makes people superstitious, angry, and metaphysical. They can't accept the split. They demand symmetry: "If I can see it, I should be able to stop it."

But symmetry is a human wish, not a property of channels.

A radio can receive a distant broadcast with a long wire and sensitive electronics, while being unable to transmit back because its power output is weak. The reception can be beautiful. The transmission can be nonexistent. No moral failure is required for this asymmetry to exist. It's just physics.

Now make it operational again with a scenario that hurts.

A remote operator controls a machine over a link with delay: a drone in wind, a rover on gravel, a crane over a crowded site, a robotic tool in a delicate procedure. At short delay, the operator can place the tool, avoid obstacles, stabilize motion. Influence is real and detectable. It clears calibrated baselines.

As delay increases, the operator begins correcting the past. The machine has already drifted. Corrections land late. The operator compensates by acting more. More action becomes more self-interference. Oscillation appears. The machine still "moves in response," but the desired placement accuracy collapses.

Here is where naive minds get fooled: they equate "responsiveness" with "control." The machine is responding to their inputs in the immediate sense—the display changes, the motors whine, the video shows motion—so they feel connected.

But connection at the input layer is not the same as steering the downstream state distribution you care about.

When calibrated against honest baselines, the operator's inputs may no longer shift the distribution of later placement outcomes beyond what the medium would produce anyway. In other words: the operator is still moving the wheel, but the world is no longer reading it as a meaningful message.

And yet the machine may still avoid catastrophe for a while, because it has its own passive stability, because obstacles are sparse, because the wind is mild, because the environment is temporarily forgiving.

That's the gap again: beyond demonstrable influence, before visible failure.

Now zoom out from machines to social processes, because the same structure repeats at larger scales.

A person can predict a cascade in a market, a supply chain, a crowd, an epidemic, a credit cycle, a power grid under stress, a climate-related extreme event pattern. Prediction remains possible because the process has inertia and slow modes; the early indicators exist; the signals are strong.

But steering—meaningfully changing the downstream distribution—may be beyond reach at the relevant delays because action bandwidth is tiny relative to the process scale, because feedback is stale, because noise is enormous, because coordination costs smear intent into incoherence.

So you get the modern curse: excellent forecasts and thin leverage.

This is the environment in which people become addicted to "early warnings" the way earlier cultures were addicted to prophecies. The warnings keep arriving. The outcomes keep arriving too. The existence of warning becomes a substitute for influence. People confuse the ability to announce with the ability to steer.

A calibrated boundary prevents that confusion.

It forces a separation between two questions:

- "Can I say what will happen?"
- "Can I make it not happen?"

It allows the first to be yes while the second is no, without turning the situation into moral melodrama.

Now, in the gap between detectability loss and functional breakdown, there's one more trap: the illusion of *micro-control.*

Even when your influence is below the floor at long delays, you may still have influence at very short delays—over the readout layer, over local metrics, over scores, over labels, over immediate responses. You can still make dials move. You can still "improve" what is being reported. You can still create immediate evidence of "doing something."

That can feel like rescue, especially when fear rises. But it can also be a way of hiding the true loss of downstream reach. You may be perfecting the display while the underlying process continues on its path.

So the boundary you're mapping is not merely a scientific curiosity. It is a shield against being seduced by short-delay compliance while long-delay steering has already vanished.

And here's the last conceptual move that makes everything click:

The detectability boundary is where your **message stops being readable** to the future at that delay.

Functional breakdown is where the **world stops being forgiving** enough to coast without readable messages.

Those are different events. Different causes. Different time scales.

Which means a process can remain "stable" while already beyond steerability in a demonstrable sense, and then fail abruptly when the environment shifts regime, when damping weakens, when noise spikes, when constraints tighten, when time pressure increases.

This is why cliffs feel like betrayal. People experience them as sudden moral reversals: "It was working, then it wasn't." But the coupling may have been gone long before. The curve arrived later.

A calibrated boundary lets you say, calmly, without superstition:

"It looked like control because conditions were easy. The medium stopped carrying me earlier. The curve is what revealed it."

That statement is not comfort. It is clarity.

And clarity is the only thing that lets you do the next necessary step: stop treating the boundary as a tragic endpoint and start treating it as a map you can move by changing noise, delay, feedback quality, and action bandwidth.

The boundary is not the end of action.

That's the cruel joke.

You can cross the line and still act all day. You can cross it and still speak, click, push, threaten, plead, optimize, hustle, "take responsibility," "show leadership," "do the work." The muscles still contract. The mind still burns. The world still accepts your inputs the way a wall accepts a fist. So if you define the boundary by how it feels, you'll never find it. It doesn't feel like a switch turning off. It feels like continued motion with thinning consequence.

That's why the only honest way to locate it is comparative: you have to watch the future and ask whether it behaves differently when you act than it behaves in null worlds where the action thread is deliberately severed but the world's structure is preserved.

And once you've done that—once you've chosen honest nulls, a tolerance bar, and a persistence rule—the boundary becomes a coordinate. It stops being a metaphor and becomes the earliest delay where your influence collapses into baseline shimmer and stays there most of the time.

But what does that coordinate actually buy you?

It buys you the right to stop lying.

It buys you a separation between two kinds of sentences that people constantly confuse:

- "I can see what's coming."
- "I can change what's coming."

Those sentences can be simultaneously true and false in different combinations, depending on delay, noise, feedback quality, and bandwidth. The boundary you're mapping is precisely the place where those two sentences split.

And the split has a strange consequence: it changes what "responsibility" even means.

If your influence at a certain delay is not distinguishable from null, then demanding that you "should have changed it" is like demanding that a dim flashlight should have carried its blink pattern through fog thick enough to smear it into shimmer. It's not a moral demand, it's a category error. It's assigning blame to a channel that cannot carry the signal under the conditions that were actually present.

This does not absolve anyone of everything. It does something more useful: it forces responsibility to track the parts of the world that still have readable transmission.

It forces responsibility to ask: who controls noise? Who controls delay? Who controls feedback quality? Who controls the record? Who decides what gets measured and when it is returned? Who has bandwidth? Who can intervene in ways that actually clear the bar?

Because if agency is a channel, then the most powerful act is often not "try harder," but "change the medium."

And this is the point where the boundary stops being merely descriptive and becomes diagnostic.

When you cross the boundary, you can stop framing your situation as a personal deficiency and start framing it as a medium problem:

- Is the environment injecting too much variability?
- Is feedback arriving too late to be useful?
- Is the record you condition on too degraded to separate shared-state drift from action effect?
- Are you overdriving the loop with frantic micro-actions that add noise rather than meaning?
- Are you stuck optimizing the readout layer while the underlying process continues unchanged?

Those questions are not comforting, but they are actionable in a way "be stronger" is not.

Now, there is a danger here too, and it's worth naming because people will exploit it: the boundary can be used as a weapon.

If I want to reduce your usable influence without touching you physically, I don't have to forbid you from acting. I just have to make sure your actions don't clear the bar.

I can do that by:

- adding noise (flood the environment with variability and distraction)
- adding delay (slow down feedback until you correct the past)
- degrading the record (reduce what you can see, fragment the context)
- forcing you into high-frequency interventions (make you "engage" constantly)
- changing what counts as an outcome (replace real downstream state with a meter you can manipulate)

Under those conditions, you will still "act." You will even act more. You will look busy. You will feel responsible. And yet your action thread will not transmit into the future in a way you can demonstrate.

This is a particularly modern form of control: not stopping you, but making your influence unreadable.

Which is why defining the boundary with calibrated nulls is not academic. It is a defense against being turned into a perpetual motion machine of futile effort.

Because once you know the boundary exists, you can stop mistaking frantic movement for leverage. You can stop worshiping isolated spikes. You can stop living in the casino of "maybe it worked this time." You can treat glitter as glitter.

And now comes the final turn, the one that should land with a quiet punch:

The boundary is not a claim about the universe. It is a claim about what can be **demonstrated** under constraints.

It says: given the record you had, given the medium you were embedded in, given your action bandwidth, and given the tolerance you set for being fooled by baseline shimmer, there is a delay beyond which your influence is not distinguishable from null.

That is not metaphysical doom. It is a measurement statement. And because it is a measurement statement, it can be tested, stressed, moved, and argued about with evidence rather than with faith. You stop asking whether you are "free."

You start asking where your transmission remains readable, and where it dissolves into fog bloom. You accept that the world can remain legible while becoming unsteerable at the ranges you care about. You accept that stability can persist after coupling is gone, and that cliffs arrive when forgiveness ends. You accept that the right way to treat a boundary is not as a tragedy but as a map: a map of where influence exists above baseline shimmer and where it doesn't.

And that map becomes the bridge to what comes next.

Because once you can locate the boundary, you can do something that sounds like it should be impossible but is actually inevitable:

You can watch prediction persist past the boundary.

You can watch the future stay readable even after steering becomes indistinguishable from null. You can watch the two curves separate—forecasting strength and steering strength—until they form a gap. That gap is the next creature you have to name.

It is the regime where you can see what's coming, but you cannot demonstrably change it at the relevant delays.

It is the condition that produces the modern experience of being correct without being effective. It is the place where early warnings multiply while interventions shrink.

It is the place where knowledge becomes a kind of spectator sport.

And once you name that place, you can stop confusing it with failure of character and start treating it as what it is: a measurable decoupling between two channels—one that brings the world into you, and one that lets you push yourself into the world.

That decoupling is not the end. It is the setup.

Because now you're ready to talk about what happens when prediction stays alive but influence dies—and why delayed steering collapses into failure cliffs even when forecasts remain sharp. That's where the next turn begins.

FORMULA DECK (Chapter 5)

The Agency Horizon, Null Calibrated Boundary of Influence

1. Setup and post zero steering curve. Use the Chapter 4 objects, spacetime as (M, g), a bounded region V inside M with boundary ∂V, and an admissible history set H on V. Let the observer have an accessible record up to time t, and let F_t be the information set generated by that record. Let an action be chosen at time t under an access constrained policy. Let an outcome be evaluated at a nonzero delay τ. Define the post zero steering curve as the measured action to outcome coupling across delays τ greater than zero. This is the quantity whose disappearance defines the horizon.
2. Null procedures that preserve structure. To avoid mistaking structured drift or shared state artifacts for agency, define one or more null procedures that break action to outcome linkage while preserving relevant temporal structure. Examples include stratified shuffles within information strata and autocorrelation preserving time shifts. For each delay τ, compute a null baseline for the steering curve using these procedures.
3. Null threshold for false alarm control. Choose a conservative calibration level β near 1, for example 0.99. For each delay τ, define a per delay null threshold as an upper tail value of the null

baseline at that delay. If the measured steering curve at delay τ is at or below this threshold, influence at that delay is not distinguishable from null under the chosen calibration.

4. Statistical agency horizon as first loss of distinguishability. Define the agency horizon as the earliest nonzero delay where the measured steering curve falls to the calibrated null threshold or below. This identifies the first delay where usable influence is no longer supported by the data under the specified null family and calibration level.
5. Post zero agency horizon as the operational definition. Because τ equal to zero is excluded by hygiene, define the operational horizon explicitly as the first delay τ greater than zero where steering becomes indistinguishable from null. Beyond this post zero horizon, attempted steering is statistically indistinguishable from the calibrated null floor given the observers accessible information.
6. Persistence based horizon for robustness. Single delay crossings can occur due to estimator noise. To reduce this, impose a persistence rule. Choose integers m and n with m at most n. Declare a delay τ to be null consistent if at least m of the next n delays are at or below their null thresholds. Define the robust horizon as the earliest delay where this null consistency condition holds. This yields a more stable horizon estimate in finite

samples.

7. Functional horizon versus statistical horizon. The agency horizon is a statistical boundary; it marks loss of detectable action to outcome coupling above null. Separately define a functional failure horizon used later, it marks practical breakdown by a chosen criterion such as failure rate exceeding a high level. The statistical horizon and functional horizon can coincide or separate depending on regime, and the gap between them is itself an empirical signature.

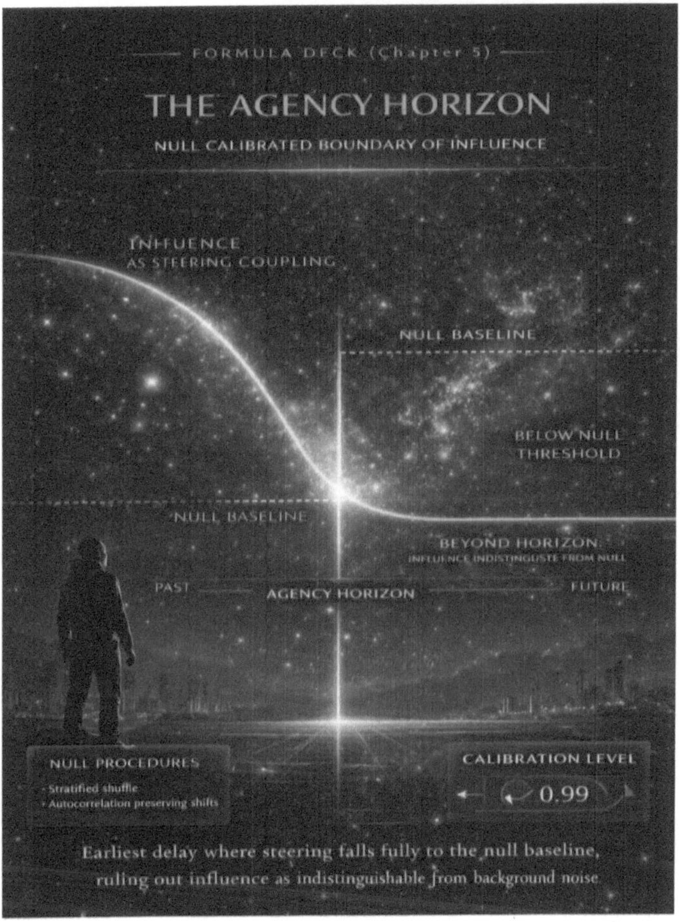

Earliest delay where steering falls fully to the null baseline, ruling out influence as indistinguishable from background noise

Chapter 6

Steering Under Delay

Imagine you are driving at night on a road you know well. Your hands rest lightly on the wheel, your eyes scan the lane, your body makes tiny corrections without drama. This feels like control because it is control. The road curves, you respond, the car turns, the world answers you quickly enough that your next correction still matters. The loop is short. You act, the world replies, you act again.

Now imagine the same road, the same car, the same hands, but every response arrives late. You turn the wheel and nothing happens for a moment. Then the car turns too much. You correct, and that correction arrives late too. You begin to overshoot. You begin to chase your own mistakes. The road has not become evil. You have not become stupid. One variable changed, delay.

Delay turns steering into a different kind of activity. It turns smooth control into a struggle against your own past actions. It forces you to drive the present with information from the past. It makes you guess rather than respond. It makes prediction feel necessary, not as a luxury, as a crutch.

This chapter is about that shift, the moment when the world stops answering fast enough, and control begins to collapse.

To understand steering under delay, start with the simplest idea. Control is not a command. Control is a conversation. You do something, the world does something back, and you adjust. That is the only mechanism. The conversation can be gentle and stable, or chaotic and unstable, but it always depends on timing.

A thermostat is a conversation. You set a temperature. The heater warms the room. The thermometer senses the warmth. The heater turns off. The room cools. The heater turns on. That loop is control. If the thermometer responds instantly and the room changes slowly, the loop is easy. If the thermometer is slow or the room changes rapidly, the loop becomes jumpy. The heater overshoots, then undershoots, then overshoots again. You are not getting warmth, you are getting waves.

A spacecraft is a conversation too, only the delays are extreme. When engineers send a command to a probe near a distant planet, they do not steer it the way you steer a bicycle. They cannot. The signal takes minutes or hours to arrive. By the time the probe receives the command, it is no longer where it was when the command was sent. The probe is living in a different now.

So what do engineers do. They move the loop. They put part of the control onboard. They give the probe rules that let it respond locally, because local response is the only way to keep the conversation alive. They accept that long distance steering is mostly planning, not control. They accept that late commands are guidance, not grip.

Human life is moving into a similar geometry. Not because you are traveling to Jupiter, because the loops around you are getting longer, noisier, and more crowded. The world answers slower, and the number of signals competing for your attention grows. The conversation that used to feel direct begins to feel like a delayed echo.

This is why so many modern people describe a strange sensation. They feel informed, yet unable to change outcomes. They feel watched, yet unheard. They feel as if the future is being shown to them, while the levers that could have mattered were earlier, smaller, and already gone.

This is not a moral diagnosis. It is a control problem.

Control has three enemies. Delay, noise, and limited energy. Delay is the most brutal because it attacks the structure of the loop itself. When feedback arrives late, you are always steering a version of the world that no longer exists. You are adjusting based on stale information. Even if the information is accurate, it is not timely. Accuracy without timing is a beautiful report that arrives after the door closes.

Noise makes the problem worse. If your measurements are fuzzy, you cannot tell whether a small change is real. You hesitate. You average. You wait for confirmation. That waiting is another delay. So noise creates delay, and delay amplifies the effect of noise.

Then there is energy. Control requires attention. It requires time. It requires the ability to notice a signal and translate it into action. If you are exhausted, your effective delay grows. The signal arrives, but meaning arrives later. Your loop lengthens from the inside.

Now, there is a common fantasy in modern culture. The fantasy says that better prediction solves delay. If we can forecast earlier, we can act earlier, and control returns. That is partly true in special cases, but it hides a deeper truth. Prediction does not shorten the loop. Prediction only tries to compensate for a loop that is already long.

Imagine driving again, but now with a map that tells you the road will curve ahead. The map is useful. It lets you prepare. But if the steering response is delayed, the map does not fix the delay. It only makes the delay survivable for longer. You are still steering with late feedback. You are still chasing your own past. You are simply guessing better.

In other words, prediction is a prosthetic for damaged coupling. It is not a replacement for coupling.

This is why some domains can become very good at forecasting and still remain hard to steer. The forecast gets better, the loop stays long, and the person experiences the gap as a kind of late clarity. They see, but they cannot move the world in time.

The core idea of the Volumetric Time Model is that leverage is local. Not local in space alone, local in time. Leverage lives in short loops. Short loops are the regions where action and consequence remain coupled strongly enough to be distinguishishable from background drift.

When the loop stretches, two things happen at once. First, you need more force to steer, because the world has more time to drift before your correction lands. Second, you become more prone to instability, because you are correcting based on old data. You push too hard because you do not see the new state yet. Then the correction lands, and it is too much. Then you correct again, and again it lands late. The result is oscillation.

This is not a metaphor. It is a universal feature of delayed feedback, whether you are heating a room, landing a drone, managing a supply chain, or trying to repair a relationship. When response is delayed, overcorrection becomes likely. Overcorrection creates volatility. Volatility increases noise. Noise increases hesitation. Hesitation increases delay. The loop becomes longer from every direction.

People often interpret this as chaos in the world. But the deeper cause is the geometry of the loop.

So what do we do with this. The first step is to stop describing control as a property of your will. Control is a property of a loop. Your will is only one component. The timing of signals, the clarity of meaning, the availability of a lever, the speed of response, the cost of attention, all of these determine whether control is possible.

The second step is to name a practical consequence. In a delayed world, the safest behavior is not always stronger action. Sometimes stronger action is exactly what produces instability. In delayed control, gentler and earlier often beats dramatic and late. A small correction applied before the drift grows can do more than a large correction applied after the drift has already carried you away.

That is why the most powerful levers in life often look boring. They are early levers. They are small levers. They are reversible levers. They are levers that keep the loop short.

Sleep is a lever because it preserves bandwidth. A clear calendar is a lever because it reduces cognitive delay. A simple habit is a lever because it creates rapid feedback. A direct conversation is a lever because it prevents long silent drift. A reversible choice is a lever because it keeps more futures reachable.

None of these levers feel heroic. They feel mundane. But mundane is often where control lives.

Now consider the opposite, the levers that look powerful but arrive late. A dramatic intervention after a long decline. A sudden policy after a slow buildup. A last minute promise after months of silence. A big button that appears after the branch has closed. These actions can be sincere, even desperate. But delayed loops punish desperation. The action lands when the world has already moved.

This is the moment when people begin to rely on prediction as comfort. They watch dashboards. They read trends. They consume forecasts. They feel clarity. But clarity is not coupling. Clarity is not a lever. There is a cruel feature of delayed control that surprises people the first time they meet it. A loop can look calm right up until it fails. It can even look improved. More data, more monitoring, more confidence. Then one day it begins to wobble, and the wobble grows, and the corrections become desperate, and the outcome becomes predictable in the worst way.

This is not because the world suddenly changed its personality. It is because a delayed loop hides its own instability until the moment it can no longer hide it.

Think again of the thermostat. In a small room with a fast sensor, the heating is smooth. In a larger room with a slow sensor, the heating becomes waves. At first the waves are small. You might not notice them. Then a few small changes occur. The outside temperature drops, the room loses heat faster, the sensor lags a bit more, and suddenly the waves amplify. The room becomes alternately too cold and too hot.

What changed. Not the intention of the thermostat. Not the basic physics. The timing margin got eaten away. The loop was stable only because it had slack. When the slack disappeared, the same controller became a generator of oscillation.

That is the first anatomy lesson. Stability often depends on slack. Slack is time margin. Slack is spare energy. Slack is clarity. Slack is the ability to absorb disturbance without overreacting. When slack shrinks, delayed control becomes fragile.

Now bring the same lesson into human life. A person can manage a busy week when they have sleep, food, and quiet moments. That is slack. Remove sleep, add noise, add uncertainty, and suddenly the person becomes reactive. They overcorrect. They answer too fast, then they withdraw, then they answer too fast again. The same personality looks different. But the deeper truth is that the internal loop lost slack.

In organizations and societies, the same pattern scales. When supply lines are loose and inventories exist, a disruption can be absorbed. When everything is tuned for minimum buffer, a disruption creates panic ordering, and panic ordering amplifies demand swings, and demand swings produce shortages, and shortages create more panic. The loop becomes a machine for its own instability.

The delayed collapse is often not a single mistake. It is a sequence of well meaning corrections applied too late.

This is why delayed steering is so difficult psychologically. The mind expects that effort should help. In many parts of life, effort does help. In delayed loops, effort can make it worse. A late correction tends to overshoot. Overshoot creates new error. New error demands new correction. The person becomes busy, but not effective.

Here is a second anatomy lesson. In delayed loops, the path to stability is often earlier action, not stronger action.

When the response arrives late, force becomes dangerous. The safest correction is the one applied before the drift becomes visible to your emotions. That is a strange sentence, but it is true. By the time your emotions are fully engaged, the loop is often already long. You are no longer steering the present. You are fighting the past.

This is why so much modern life feels like fighting the past. People do not feel late because they are lazy. They feel late because the feedback they receive is often a polished summary of what has already happened. When the summary arrives, the only available levers are large and blunt. Large and blunt levers create collateral effects, which create more noise, which further delays meaning.

Delayed control has another signature that engineers know well and ordinary people experience without naming. The signature is chasing.

Chasing means you aim your actions at a world that is already gone. You see a number rise, you react, the reaction lands after the number has already changed, and you react again. Many market panics are chasing. Many relationship arguments are chasing. Many health crashes are chasing. A delayed loop trains you to chase because your measurement arrives after the state moved.

The third anatomy lesson is about measurement itself. More measurement does not guarantee more control.

This is hard for modern minds because we live in a culture that worships observation. We imagine that if we can see enough, we can steer enough. The universe teaches a harsher rule. Seeing and steering are different couplings.

You can build a camera that sees the curve ahead. But if the steering linkage is loose, the camera does not fix the linkage. You can build a forecast that is nearly perfect. But if the lever is not available inside the effective window, the forecast becomes a beautiful report.

In delayed loops, measurement can even reduce control by creating temptation. The temptation is to act on every fluctuation. When you act on every fluctuation with delayed feedback, you create oscillation. You create noise with your own hands.

This is why some of the most stable controllers are calm. They do not react to every twitch. They react to trends, and they do so gently, and they do so early. The goal is not to be busy. The goal is to keep the loop from leaving its stable region.

Now consider the most modern kind of delay. Not physical distance. Not slow sensors. The modern delay is gating.

A lever can exist, but it can be placed behind friction. Extra steps. Extra confirmations. Extra waiting. In many areas of life, people are offered visibility first and action later. They are shown charts, risk scores, rankings, warnings. Then action is routed through slow channels.

This creates a strange emotional split. The person knows, but cannot act. The person predicts, but cannot steer. The person receives clarity as a kind of burden.

When levers are delayed, two things happen. First, the effective window shrinks. Second, people adapt by turning prediction into ritual. They watch more. They refresh more. They read more. They hope that more observation will create leverage. It usually does not. It often creates exhaustion, and exhaustion lengthens the loop from the inside.

That is how a culture can become saturated with signals and still feel helpless. The beam gets brighter, the handles get farther away.

In control language, you can say it simply. The observational channel improved. The action channel did not.

Now we can describe delayed collapse in a way that feels almost like a law of nature.

When a loop is long, stability requires small early corrections. If small early corrections are blocked, people will wait. When people wait, drift accumulates. When drift accumulates, only large corrections remain. Large corrections applied with delay overshoot. Overshoot creates new drift. The loop enters oscillation. Oscillation increases noise. Noise increases hesitation. Hesitation increases delay.

This is a spiral. It is not destiny. It is a predictable geometry.

The good news is that geometry suggests remedies that are surprisingly practical. Not grand. Not heroic. Practical.

The first remedy is to shorten loops wherever possible. This can mean moving decision making closer to where the consequences appear. It can mean automating reversible actions that keep the state within safe bounds. It can mean giving people direct levers earlier instead of giving them perfect warnings later.

The second remedy is to increase reversibility. Reversibility is how you keep oscillation from becoming catastrophic. When errors can be undone, early action becomes safer. When errors cannot be undone, people become cautious. Caution delays action. Delayed action forces large corrections. Large corrections are unstable under delay.

Reversibility is not softness. It is stability engineering for human beings.

The third remedy is to protect signal integrity. If every message looks urgent, then urgency loses shape. If signals arrive as a flood, then detection becomes late. Late detection shortens the effective window. The loop lengthens. Control collapses.

So the remedy is not more alerts. It is fewer sharper signals. Clearer meanings. Less clutter.

The fourth remedy is to respect energy. A person with depleted energy is a controller with increased internal delay. The signal arrives, but recognition arrives later. Recognition arrives, but action arrives later. That extra delay can be the difference between a usable window and a closed door.

This is why sleep and rest are not lifestyle advice inside this model. They are access support. They are what keeps your cone wide enough to overlap with branching.

Now let us make the cosmic part of this chapter visible.

A planet orbits a star in a stable pattern because feedback is local and immediate. Gravity acts continuously. There is no waiting for permission. There is no delayed approval. The forces are there, all the time, and the orbit is a conversation between inertia and attraction.

Human life is not an orbit, but the metaphor teaches a simple truth. Stability belongs to continuous local coupling. When coupling is made discrete and delayed, stability becomes harder.

A world built out of delayed prompts and late levers will trend toward chasing and oscillation. It will produce people who feel as if they are always catching up. It will produce institutions that act only when the dashboard turns red, then act too hard, then calm down, then repeat.

This is what makes forecasting without power such a natural attractor. Prediction improves, because more sensors exist. Steering collapses, because action remains delayed, gated, and expensive.

So the question this chapter forces is not whether prediction is good. Prediction can be wonderful. It is one of the great achievements of science. The question is whether prediction is matched with reachable levers. Whether windows remain open. Whether correction can happen early. Whether action can be gentle rather than desperate.

You can now see why a warning that arrives with no lever is not a gift. It is a timing receipt disguised as help. It tells you what will happen, and then it places you in the psychological position of being responsible for what you cannot reach.

That is not only unfair. It is unstable.

In the next part we turn from anatomy to design. We will map the minimal features of a steerable world. Not a perfect world. A steerable one. A world where the loop remains short enough that human action is still distinguishable from noise. A world where the future can be readable without becoming a prison.

A steerable world is not a world with perfect forecasts. A steerable world is a world where the loop stays short enough that action still changes outcomes in time. If you want a single picture to hold in your mind, it is this. Your hand moves, the world answers, your hand moves again. The answer arrives soon enough to matter.

When that answer arrives late, people try to compensate by staring harder. They collect more information. They refresh more screens. They demand more certainty. They treat prediction as a replacement for response. That is how a culture can become obsessed with dashboards and still feel powerless. The eye becomes stronger, the grip becomes weaker.

So the design question is simple. Not easy, simple. Where do you place the loop.

In engineering, whenever delay grows, the only stable move is to relocate control closer to the thing being controlled. A distant pilot cannot fly a drone smoothly if the video feed is delayed. The answer is not a better video feed alone. The answer is local autopilot, local stabilization, local correction. The distant pilot becomes a planner, not a steering wheel. The loop is moved back toward the machine, where the response is fast.

Human life has the same logic. If you want a person to steer, you have to place levers where their actions can close a loop quickly. If the lever exists only after the branch closes, then the lever is theater. If the lever is buried behind delay, then the lever is functionally absent.

This is the first principle of preserving agency under delay. Put the lever upstream.

Upstream does not mean earlier in a calendar sense only. It means earlier in the causal chain. Earlier in the sequence where small actions still compound. Earlier where reversibility is still high. Earlier where gentle correction is possible.

A second principle follows immediately. Make actions continuous when possible.

Many modern levers are discrete. A large irreversible decision that arrives after a slow buildup. A yes or no choice presented at the last moment. A heavy intervention that lands after the drift has become dangerous. Discrete late control tends to produce oscillation, because it forces large corrections when the loop is already long.

Continuous control feels boring. It feels like tiny adjustments. It feels like gradual correction. It feels like daily habits and quiet maintenance. But continuous control is how stable motion happens in nature. Gravity is continuous. Friction is continuous. A planet does not receive a weekly permission slip to remain in orbit. The coupling is always there.

When you design human levers to be continuous, you reduce the need for dramatic late actions. You let people correct early without panic. You make small changes safer, because the cost of error is smaller.

This connects to the third principle. Increase reversibility.

Reversibility is the safety valve of control. If you can undo an error, you can act earlier. If you cannot undo an error, you hesitate. If you hesitate, the loop lengthens. If the loop lengthens, only large corrections remain. Large corrections under delay overshoot. Overshoot creates new error.

Reversibility changes this spiral. It lets early action become rational. It keeps people from waiting until the moment they are forced. It keeps the corridor wide for longer.

Reversibility can be literal, like an undo window for a financial transfer, or a cooling period for a major commitment, or a chance to reverse a choice before consequences lock in. Reversibility can also be social, like a culture where apology is possible early, where repair is rewarded, where small misunderstandings are corrected before they harden into permanent stories.

A steerable environment makes early repair easy. A brittle environment makes repair humiliating, expensive, or delayed, and then acts surprised when everything breaks.

The fourth principle is signal integrity.

A control loop depends on detecting the right signal at the right time. If the signal is buried in noise, then it arrives late in practice, even if it arrives early on paper. People do not respond to what they technically receive. They respond to what they detect, understand, and believe.

A culture of constant prompts is a culture that damages signal integrity. When everything is urgent, nothing is urgent. When every message demands attention, attention becomes numb. When the mind is flooded, meaning arrives late. Late meaning shortens the effective window.

So signal integrity is not about making everything louder. It is about making the few critical things distinct. It is about reducing clutter. It is about giving warnings a shape that the brain can recognize quickly. It is about using plain language. It is about not hiding consequence behind soft phrasing.

Plain meaning is part of timing.

If a sentence takes ten seconds to understand, then you lost ten seconds of the window. If a screen requires multiple steps to interpret, then you lost more time. If the prompt invites confusion, then confusion becomes delay, and delay becomes collapse.

The fifth principle is energy honesty.

A human being is a physical device with finite bandwidth. If you ignore that, you create loops that look fine on paper but fail in the real world. Sleep loss increases reaction time. Stress narrows attention. Overload increases hesitation. Hesitation increases delay. This is not psychology as opinion. This is biology as physics.

A steerable environment respects the energy budget. It does not demand constant vigilance as the price of safety. It does not punish people for needing rest. It does not require a perfect mind to avoid disaster.

If safety requires perfect attention at all times, then safety is fake.

So the environment must be designed so that ordinary human energy is enough. Otherwise, the loop fails in the predictable way it always fails, slowly at first, then suddenly, then irreversibly.

Now there is one more idea that ties these principles together, and it is simple enough to fit on a small piece of paper. Make timing visible.

Most people do not need more moral lectures. They need a receipt for when the window was open. They need a clear record of when the signal arrived, when the lever became available, and when the branch closed.

When timing is hidden, blame becomes cheap. A person is told they should have acted, without anyone proving that action was possible in time. A person is told they were informed, as if being informed is the same as being empowered.

When timing is visible, the conversation becomes honest. If the signal arrived late, it is clear. If the lever arrived late, it is clear. If the door closed early, it is clear. And then design can improve, because the failure is measurable.

Timing visibility does not require storing personal content. It requires timestamps. It requires minimal records. It requires standards that say, for high stakes outcomes, the effective window must be positive, reversibility must exist for long enough, and levers must be real in the only sense that matters, actions must couple to outcomes above background drift.

When timing is visible, a culture learns the difference between a warning and a lever.

This matters because there is a seductive trap in the modern age. The trap is that visibility feels like fairness. If you can show someone a forecast, you can claim you did your job. If you can present a risk score, you can claim transparency. If you can display a chart, you can claim responsibility moved to the user.

But visibility without leverage is not fairness. It is often the opposite. It hands a person foreknowledge without tools. It creates guilt without control. It converts the pain of delayed control into personal shame.

A steerable world refuses that seduction. It does not treat a forecast as a substitute for a lever. It treats forecasts as valuable only when they arrive inside usable windows, attached to reachable actions, supported by reversibility, protected by clean signals, and calibrated to human energy.

If you want a final image to carry from this, return to the night driving scene. A road is safe not because you have a perfect map of every curve. A road is safe because your eyes can see far enough, your hands can move the wheel, the tires respond, and the car turns before the cliff.

Add more headlights, yes, but also keep the steering linkage tight. Keep the brakes responsive. Keep the road markings clear. Keep the guardrails in place. Make errors undoable. Do not demand that the driver predict every gust of wind perfectly.

That is what it means to preserve agency under delay. It is not about forcing freedom into a finished universe. It is about protecting the local conditions where action still matters. It is about keeping coupling alive.

When coupling is alive, even a small person can change their trajectory in meaningful ways. When coupling is dead, even perfect knowledge becomes a spectator sport.

In the world ahead, prediction will keep improving. That part is almost guaranteed. The harder question is whether leverage will be protected at the same rate. If it is not, then many domains will drift toward a strange stable misery. Accurate forecasts, late levers, narrow windows, high confidence, low control.

If leverage is protected, the future will feel less obvious. That is the paradox. More steerability means more real branching, which means more uncertainty, which means less of the cold late clarity that feels like punishment.

Uncertainty is not always ignorance. Sometimes uncertainty is the signature of freedom that still exists.

FORMULA DECK (Chapter 6)

Steering Under Delay, Loop Length, and Control Collapse

1. Delayed dynamics. Consider a process whose underlying state evolves over time. Actions chosen by an observer do not take effect immediately. Instead, actions arrive after a delay measured in discrete steps. This makes explicit that interventions can land late relative to the evolving state, even when the action itself is correct.
2. Delayed observation and noisy access.
   The observer does not see the current state directly. Observations arrive after a sensing delay and are corrupted by noise. The observer's present picture is therefore always a delayed and imperfect view of what is actually happening.
3. Information record and monotone access.
   Let the accessible record be the accumulated data available to the observer up to a given time. As time advances, this record grows monotonically. Decisions are always made using what has arrived so far, not what exists globally in the underlying process.
4. Feedback policy under access constraints.
   A feedback policy maps accessible information into actions. Even a perfectly designed policy is constrained by what has entered the record in time. Information that exists but has not arrived

cannot be acted upon.

5. Decision latency and total loop delay. Between receiving a signal and issuing an action, additional delay arises from internal processing, coordination, and real-world friction. The total loop delay is the sum of sensing delay, decision latency, and actuation delay. This total delay determines how far behind the process the controller is forced to operate.
6. Structural role of delay in feedback loops. Near a stable operating regime, the process can be approximated by a linear response where delayed inputs and delayed observations enter explicitly. In this form, delay is not an afterthought or implementation detail. It is a structural element of the feedback loop.
7. Bounded levers and limited authority. Real interventions are constrained. Actions have limited magnitude and limited authority. Even when the future is readable, bounded action power can prevent meaningful steering when drift is strong, disturbances are large, or delays are long.
8. Readability under delay. Readability measures how much information current observations carry about future states. Large readability means the future remains statistically predictable from what the observer can see, even in the presence of delay and noise.
9. Influence under delay.

Influence measures whether chosen actions measurably change future outcomes beyond what current access already implies. Influence requires not only correct intent, but timely arrival. Delay can destroy influence while leaving readability intact.

10. Null controls and apparent influence.
Apparent influence can arise even when true control is absent, due to shared drift, estimator bias, or coincidental alignment. Null controls that break action–outcome alignment while preserving structure define a fake influence floor. True steering must rise above this floor to be operationally meaningful.
11. Control collapse under increasing loop delay.
As total loop delay grows, steering collapses before prediction does. There exists a regime where the future remains readable but actions no longer couple to outcomes in a detectable way. This is the structural origin of control collapse under delay.

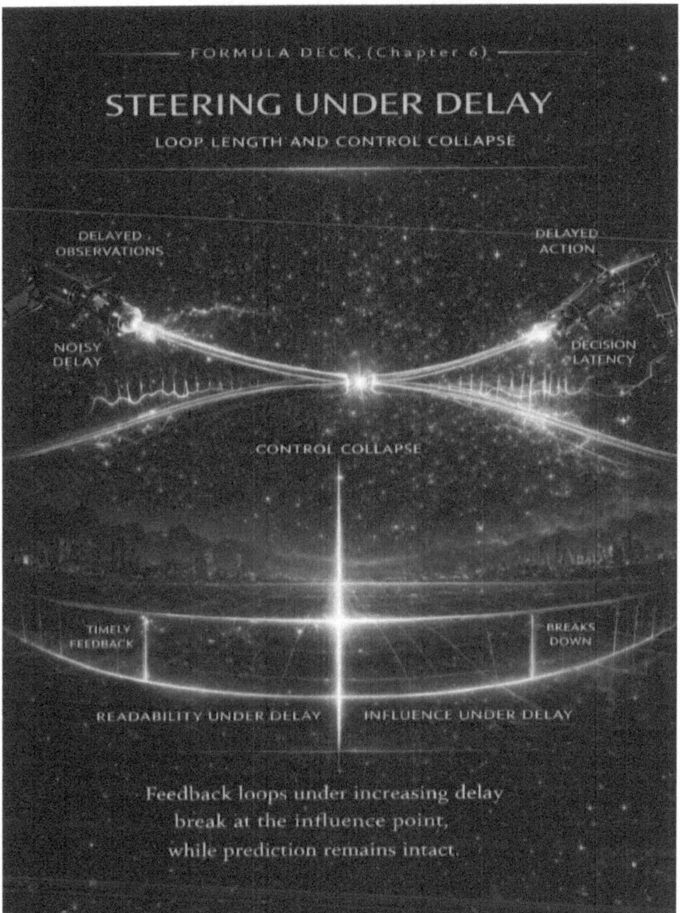
FORMULA DECK (Chapter 6)
STEERING UNDER DELAY
LOOP LENGTH AND CONTROL COLLAPSE
DELAYED OBSERVATIONS
DELAYED ACTION
NOISY DELAY
DECISION LATENCY
CONTROL COLLAPSE
TIMELY FEEDBACK
BREAKS DOWN
READABILITY UNDER DELAY
INFLUENCE UNDER DELAY
Feedback loops under increasing delay
break at the influence point,
while prediction remains intact.

Chapter 7

Forecasting Without Power

There is a strange modern experience that would have sounded like science fiction to your grandparents. You can know what is coming and still be unable to change it. You can see the wave forming and still be standing in the same place when it breaks. Not because you are lazy. Not because you are stupid. Not because reality is mocking you. Because the world can become readable at the exact moment it becomes hard to steer.

Physics gives us the cleanest metaphor for this, because physics is full of knowledge that arrives without power. When astronomers watch a distant star brighten and fade, they are watching a message that left long ago. They can measure its spectrum, estimate its mass, predict its fate, and reconstruct the story with breathtaking confidence. They cannot reach into that scene and change a single photon. Their knowledge is real. Their leverage is zero. It is not a moral failure. It is geometry plus delay.

Now shrink that cosmic picture down to the scale of an ordinary week.

A storm forms far offshore. The forecast becomes clear early. The path tightens. The models agree. The map on your screen looks crisp, almost comforting in its precision. Then the warnings arrive in a flood. Roads close. Shelves empty. Schedules collapse. People do not act earlier because acting earlier requires levers that are not actually available. Time off is not available. Money is not available.

Child care is not available. A safe place to go is not available. The forecast is correct, the future is readable, and yet the lived options are narrow. The wave makes landfall exactly where it was expected to make landfall, and many people discover that knowledge did not buy them the one thing they wanted most, a wider set of reachable exits.

Or take a quieter example, the kind that does not make the news. A bank transfer clears. A confirmation arrives. Then a fraud alert appears. You are told that something is wrong, you are told that the pattern matches a scam, you are told that you should act immediately, and by the time you are told, the action that would have mattered is already in the past. You call. You wait. You explain. You are polite because you have learned that politeness is often the only lever you still have. The outcome proceeds exactly as the alert implied it would proceed. You were informed. You were not empowered.

Or take the most intimate version of all, the slow cooling of a relationship. It does not end with a dramatic explosion. It ends with delay. Messages take longer. Plans get postponed. Repairs get deferred. The future becomes readable in the small shapes of attention, in the way a voice changes, in the way laughter thins, in the way silence becomes normal. You can see the ending coming long before it arrives. And still you do nothing that matters, because what mattered was not a grand speech at the end, it was a small early loop, a short honest conversation before the drift became habit. The future was readable, and the levers were earlier than the moment of clarity.

These examples feel different on the surface, but they share the same internal structure. A forecast arrives. The window for meaningful change is already closing, or already closed. The world gives you a view, and does not give you a grip.

This is the paradox of the modern age. Our view has improved faster than our grip.

We have built instruments that can detect faint signals, long before the consequences become visible to human senses. We can see trends in bodies, in weather, in markets, in supply chains, in crowds, in behavior, in mood. We can forecast probabilities with confidence that would have seemed miraculous a generation ago. The future becomes legible in advance, sometimes painfully legible. And yet, for many outcomes, the levers remain local, small, and time sensitive. Levers live in short loops. When the loop stretches, the lever fades.

This is the regime that needs a name, because unnamed experiences become superstition. People blame themselves. They say they should have known. They say they should have acted. They replay the moment of late clarity as if late clarity implies late guilt. But the real story is often simpler and harsher. Knowing is not the same as reaching. Seeing is not the same as steering.

Forecasting without power is what happens when readability remains high while steerability collapses. It is the experience of being shown the future with confidence, while the ability to bend that future has already fallen below the threshold where action still matters. You can still move. You can still click. You can still sign. You can still react. Yet the world no longer answers your motion in time.

Once you notice this pattern, you begin to see why it is spreading.

First, complex processes often narrow as they approach an outcome. At the beginning, many futures are possible. As time passes, options disappear. Paths merge. Constraints tighten. The corridor forms. This narrowing makes the future easier to predict, because there are fewer exits left. Prediction becomes easier precisely because the world has become less steerable. The very same narrowing that steals leverage can sharpen forecasts.

Second, more sensing does not automatically create earlier action. More sensing creates more information. But information must arrive with a usable lever. If a warning arrives early but the lever arrives late, the warning becomes a burden. It hands you foreknowledge without a handle. It can even create paralysis, because you feel responsible for what you cannot reach.

Third, human attention is a finite resource. When warnings multiply, meaning gets delayed. The signal can arrive on time and still be detected too late, because it is buried among dozens of other prompts, each demanding that you care. When everything is urgent, urgency loses shape. When urgency loses shape, the effective window shrinks in practice even if it exists on paper. The loop lengthens from the inside. The mind becomes a delayed controller.

There is an emotional signature to this regime. It is not simple fear. It is not simple sadness. It is a specific nausea, the feeling of arriving at the right answer after the moment when the answer could still change anything. It is late clarity. It is the sensation that the world is offering you information as a substitute for agency.

That is why this regime feels like punishment. Not because the universe is cruel, but because the mind expects that knowing should help. When knowing does not help, the mind turns inward. It turns into blame. It turns into shame. It turns into magical thinking. It turns into obsessive monitoring, as if staring harder will turn a view into a grip.

Obsessive monitoring is a predictable adaptation. If you cannot steer, you try to see earlier. If you cannot see earlier, you try to see more. If you cannot see more, you try to see constantly. This turns forecasting into a ritual. It does not restore leverage. It drains energy. Drained energy increases internal delay. Internal delay shrinks windows further. The spiral tightens.

A civilization can end up living inside that spiral. It can become highly informed and quietly powerless, because it invests in observation and neglects steerability. It builds better predictions and forgets to keep levers reachable. It delivers clarity and calls it fairness. It delivers warnings and calls them protection. It delivers receipts and calls them choices.

The quiet truth is that a warning is only as good as the lever that comes with it. A forecast is only as good as the window it opens. Knowledge without an actionable window is not guidance, it is exposure.

There is a more hopeful truth embedded in the same geometry. If forecasting without power is a regime, then it has boundaries. It is not a law of nature that must dominate every domain. It emerges when loops are long, when reversibility is low, when levers are delayed, when signals are drowned, when human energy is treated as infinite. Change those conditions and the regime weakens. Protect short loops and steerability returns in the places where it can return.

The first step is simply to stop confusing a view with a grip. The future can be readable and still be unreachable. If you learn that distinction, you stop treating late clarity as a personal verdict. You start treating it as a measurement. The window was too small. The lever was too late. The signal was too buried. The loop was too long.

And once you treat it as a measurement, a different question becomes possible, a question that can be answered with design rather than with guilt. Where are the real levers, and how early can they be placed, and how long can they remain reversible, and how clean can the critical signals be kept, so that knowledge becomes action before the corridor closes

The separation between readability and steerability is not philosophical, it is mechanical, and once you see the mechanics you stop being surprised by the feeling of late clarity.

Start with a simple principle, prediction is about patterns in what you can observe, steering is about changing what happens next. These are related, but they are not the same coupling.

Readability can improve for reasons that have nothing to do with your levers. You can add sensors. You can add logs. You can add higher resolution. You can add memory. You can add models that are trained on millions of past cases. All of that can make the future easier to infer from the present. It can make the map sharper.

Steerability depends on something more fragile. It depends on timing. It depends on whether an action applied now produces a distinct change later that rises above background drift, noise, and interference. It depends on whether the world answers your motion fast enough that you can correct again before the branch closes.

That is why a world can become more predictable while becoming less steerable. The observational beam improves, while the action loop stays long, costly, delayed, and crowded.

There is a second reason this separation becomes common in complex environments. Many processes become more constrained as time passes. Early on, there are many viable branches. Later on, the branches collapse into corridors. As corridors form, the outcome becomes easier to predict because there are fewer exits left. Readability rises because the future has fewer shapes remaining. At the same time, steerability falls because the same narrowing removes options.

This is why late clarity feels sharp. The mind senses the corridor even before it names it. A corridor has a distinctive emotional texture, a sense that the decision has already been made somewhere, even if no one is speaking. The practical truth is that the corridor is often made of accumulated delays. The signals were there. The levers were earlier. The window was small. By the time the corridor is visible, the only remaining actions are large and blunt.

Large and blunt actions do not restore steerability in a delayed loop. They often create new problems. They create overshoot. They create side effects. They create noise. Noise slows meaning. Slower meaning shrinks the remaining window. So even sincere action can be punished by geometry if it lands too late.

That is one of the cruel lessons of delay. The later you are, the more force you must apply, and the more that force destabilizes the world you are trying to correct. Late force is a generator of volatility.

This is the third reason forecasting without power spreads. When control becomes difficult, people compensate by monitoring. Monitoring feels like action because it produces motion in the mind. It produces updates. It produces clarity. But monitoring is not steering unless the monitoring is paired with a lever that can still move the outcome.

In many domains today the monitoring channel has been industrialized. The levers have not.

So you get the modern pattern. A person receives fine grained knowledge of risk, then is routed into slow channels for action. They are told to wait. They are told to submit. They are told to confirm. They are thanked. Meanwhile the outcome proceeds, because the branch point does not pause for human process.

Another mechanical ingredient makes this worse, attention is not infinite. When signals multiply, detection time grows. Even if a warning arrives early on paper, it can arrive late in the mind. The brain must notice it, classify it, believe it, translate it into a plan, and execute the plan. Each of those steps consumes time and energy. When the environment is flooded, the mind delays.

This creates a subtle effect. The environment can claim that warning arrived early. The person experiences it as late. Both can be true.

A flood of prompts can destroy timing without touching a clock.

Now consider what happens when an environment tries to solve this by adding even more prediction. It sounds sensible. If you warn earlier, people have more time. But if the lever remains delayed, earlier warning does not help. It can harm. It extends the duration of helpless awareness. It turns foreknowledge into stress. Stress narrows attention. Narrow attention delays meaning. Delayed meaning shrinks the effective window. The environment responds by issuing even more alerts. The cycle tightens.

This is why a culture can become highly foreseeing and highly anxious at the same time.

It sees more, and it can do less, and it feels responsible for everything it cannot reach.

Now let us walk through a few domains where the mechanics are easy to feel.

In medicine, prediction has improved dramatically. Trends can be detected early. Risk can be scored. A case can be flagged before the visible crisis. But steerability depends on staff, time, beds, authority to act, and the ability to intervene within a window when the body is still reversible.

If the warning arrives but the intervention is delayed by queues, by workflow friction, by overload, then prediction becomes a report about an approaching cliff. It can even become a moral weapon. You were warned. You should have acted. But the lever might not have been reachable.

In finance, markets price in information rapidly. That makes some outcomes more predictable. At the same time, individual actors often have levers that are slow. Credit decisions, liquidity access, emergency buffers, these do not appear on demand. So you get a world where risk becomes visible early, yet the ability to adjust position is limited by delay and cost. People can watch the future tighten, then be forced to move only when the corridor is already narrow, when the price of motion is highest.

In supply chains, sensors can show disruption earlier. But steering requires inventory, alternate routes, coordination across firms, and time margins. When everything is tuned for minimal slack, steerability collapses. The environment becomes predictable, because once one constraint binds, many downstream outcomes become almost certain. Prediction rises. Power falls.

In relationships, the instruments are not sensors and models, they are memory and pattern recognition. A person can forecast the ending because the signs are there. But steering requires short loops. It requires early repair. It requires timely honesty. If those levers are delayed, if conversations are postponed, if small cuts are not addressed, then the corridor forms. The ending becomes predictable, and at the same time it becomes hard to change, because by the time the ending is obvious, the internal state has hardened.

Across these domains, the same geometry repeats. The readable future does not imply a steerable future. There is also a deeper reason why forecasting can remain strong after leverage collapses. Prediction can piggyback on structure that is not controllable by the observer. If a process is dominated by large background forces, then the observer can infer its direction without being able to alter it. Weather is the clean example. A person can know the storm track. The person cannot move the storm. The person can only move themselves, and even that depends on levers that might be delayed.

This gives a simple diagnostic that often cuts through confusion. When you feel late clarity, ask whether the levers you needed were local, early, and small, and whether they were actually reachable when the first meaningful signals appeared. If they were not reachable, then the experience is not a personal failure, it is a coupling failure.

A second diagnostic is equally simple. If an environment gives you ever more detailed forecasts, but does not shorten the time between action and consequence, then it is investing in readability while neglecting steerability. The result is a culture of spectators, highly informed, constantly reacting, rarely able to alter the trajectory in time.

If you want the most important practical rule that emerges from this mechanical picture, it is this. Prediction is valuable only when it arrives attached to timely levers, inside a usable window, with reversibility still available, and with signals clean enough to be detected without heroic attention. Otherwise prediction is not guidance. It is exposure.

Exposure feels like responsibility. It feels like you have been handed the future, so you must be able to change it. That feeling is an illusion created by information. The world does not change because you were shown it. The world changes only when an action couples to outcomes in time.

Once you accept that, you stop demanding that clarity alone should save you. You start demanding windows. You start demanding real levers. You start demanding reversibility. You start demanding clean signals. You start protecting your own energy because energy is part of timing.

That shift is the beginning of a sane life inside a finished volume.

The pattern has a shape, and once you see the shape you stop treating it like bad luck.

There is a region of time where action still couples to outcome, and beyond that region your action becomes statistically invisible, not because you stopped moving, because the world stopped answering fast enough. At the same time, the future can remain readable, sometimes more readable than ever, because the corridor has narrowed. Fewer exits remain, so the map looks cleaner. That is the paradox. The cleaner the map, the more likely it is that the steering wheel has already gone soft.

That paradox is not an error. It is what delayed complex environments naturally drift toward, unless someone spends real effort protecting the loop.

It helps to say this out loud in a way that removes shame. Late clarity does not mean you failed. Late clarity often means you finally received a signal that was always going to arrive late, or you finally recognized a signal that was buried until your attention could no longer ignore it, or you finally reached the point where the corridor became visible because the corridor became real. Clarity at the end of a process is cheap, the world is doing most of the work for you. The narrowing itself produces predictability.

So the question is not why you did not predict earlier. The question is why you did not have a lever earlier, or why the lever was not reachable, or why the environment was designed so that early levers were hidden, delayed, or converted into ceremonies.

There is a simple way to test whether a forecast is being used as protection or as anesthesia.

If a forecast arrives with a reachable action, and that action is reversible enough to be safe, and the meaning is clear enough to be recognized quickly, then the forecast is help. It opens options.

If a forecast arrives without a reachable action, or with actions that exist only as decorative clicks, or with meaning buried under soft phrasing and clutter, then the forecast is not help. It is a receipt. It proves you were informed. It transfers blame. It extends helpless awareness. It turns the future into something you are forced to watch.

A forecast can be accurate and still be harmful, if it arrives after the last usable branch point. It can be accurate and still be unfair, if it arrives early but the only lever arrives late. It can be accurate and still be cruel, if it arrives in a flood that overwhelms attention and delays recognition until the window is gone.

The moral of the pattern is not that prediction is bad. Prediction is one of the great achievements of science and measurement. The moral is that prediction must be paired with steerability, and steerability has requirements that cannot be faked.

Steerability requires timing.

Steerability requires a lever that actually moves outcomes, not just screens.

Steerability requires a window that is positive in lived time, not only in theory.

Steerability requires reversibility, because without reversibility early action becomes too risky, and people wait, and waiting converts small corrections into late desperate corrections.

Steerability requires signal integrity, because a warning that cannot be detected is a warning that did not arrive.

Steerability requires energy honesty, because a person with depleted attention is a delayed controller, meaning arrives late even when it is delivered early.

This yields a rule that is blunt enough to be useful.

Do not treat information delivery as fairness. Treat window design as fairness.

A person is not made responsible by being told. A person is made responsible only when they had a real chance to alter the outcome, which means the signal and the lever overlapped before irreversibility, and the lever had real coupling.

Anything else is theater.

This is why the most important improvements in a high prediction world are often boring improvements. Shorter loops. Cleaner signals. Undo periods that match the stakes. Defaults that are reversible. Friction removed from urgent actions. Clear language that reveals consequence immediately. Fewer prompts with sharper meaning. Earlier placement of levers in the causal chain.

These changes do not sound like science fiction. They sound like craftsmanship. That is exactly the point. A world saturated with prediction does not need more visions of the future. It needs mechanisms that preserve the ability to act while action still matters.

There is an even deeper twist that is easy to miss. The presence of forecasting without power can make a society appear wise while becoming less free. People can point to accurate forecasts and say, look how much we know. They can treat that knowledge as proof of progress. Meanwhile the reachable options shrink quietly. The sense of agency shrinks quietly. The corridor tightens quietly. The future becomes obvious more often. People begin to confuse obviousness with inevitability. They begin to confuse inevitability with maturity. They begin to accept the anesthetic.

The antidote is to protect uncertainty where uncertainty is the signature of branching that still exists. When a domain becomes too predictable, too clean, too confident, it may be a sign that steering has already died and only one path remains. That is not always bad, some processes should be stable. But when stability is achieved by removing levers rather than by improving coupling, the calm is counterfeit.

So the right relationship to prediction is not worship and not rejection. It is discipline.

Ask one question every time a forecast appears.

What can I do that still changes this.

If the answer is nothing, then treat the forecast as a measurement of a closed window. Do not turn it into shame. Do not turn it into obsessive monitoring. Use it as evidence that the lever needs to be moved upstream next time, or that reversibility needs to be expanded, or that signal integrity needs to be restored, or that energy needs to be protected so detection is not delayed.

If the answer is something, then act early and act gently. Early and gentle beats late and violent in delayed loops. Early and reversible beats late and irreversible. Early and clear beats late and complex. The objective is not to feel in control, the objective is to keep coupling alive.

A final cosmic image can help set this in the bones.

You can watch a distant star and learn its story with astonishing precision. You cannot alter its past. But you can still alter your own local future, because your local loop is short enough to respond. You can still steer your body toward safety. You can still steer your attention toward meaning. You can still steer your relationships through early repair. You can still steer a community by building levers that are reachable in time.

The universe can be finished and your life can still contain real branching, because branching is not magic, branching is the lived experience of being inside regions where the world answers you in time. Forecasting without power is what it feels like when those regions shrink.

The only sane response is to stop mistaking forecasts for handles, and start protecting the handles themselves.

FORMULA DECK (Chapter 7)

Forecasting Without Power, The Leverage Gap, and Readability After Influence

1. Setting and admissible histories. Let spacetime be represented by (M, g). Fix a compact region $V \subset M$ with boundary ∂V, and let H denote the admissible history set on V. Let ϕ in H represent a globally consistent history inside V. An embedded observer does not access ϕ directly, but only through delayed and noisy channels.
2. Observer record and information growth. Let D_t denote the data accessible to the observer up to time t, and define the information set $F_t = \sigma(D_t)$. As time increases, access grows monotonically. For $s \leq t$ we have $F_s \subseteq F_t$. This formalizes that uncertainty evolves through access, not through creation of new histories.
3. Outcome and forecast target. Fix a future outcome of interest defined at a delay τ greater than zero. A forecasting task is to estimate the distribution of that future outcome given current access. Forecast quality can be summarized by a proper scoring rule or by information based measures that quantify how much current access reduces uncertainty about the future.
4. Readability as prediction coupling. Prediction coupling at delay τ measures how strongly the future outcome is statistically related to what the observer can currently see. Large prediction

coupling indicates that the future remains readable from present access even when effective control is absent.

5. Actions and bounded levers. Let an action be chosen under an access constrained policy, meaning the action is a function of what is accessible at decision time. Impose a bound capturing limited authority. In realistic domains, bounded authority can coexist with strong readability.
6. Influence as steering coupling. Steering coupling at delay τ measures whether chosen actions measurably influence future outcomes beyond what current access already implies. This distinguishes true leverage from mere alignment.
7. Null calibrated influence threshold. Define a null baseline for steering coupling using alignment breaking controls, for example shuffled actions and time shifted actions. Choose a conservative threshold from the upper tail of this null baseline. Declare influence present only when measured steering coupling exceeds the threshold.
8. Post zero agency horizon. Define the post zero agency horizon as the smallest nonzero delay at which steering coupling falls to the null threshold or below. Beyond this horizon, steering is operationally indistinguishable from the null floor even though actions continue to be taken.
9. Forecasting Without Power regime. The

Forecasting Without Power condition at delay τ is the conjunction of strong readability and absent influence. The FAWP region is the set of delays where prediction coupling remains substantial while steering coupling is at or below the null threshold.

10. The leverage gap. Define the leverage gap at delay τ as the difference between prediction coupling and steering coupling at that delay. In a steerable regime, prediction and steering track together and the leverage gap is small. In the FAWP regime, prediction remains large while steering collapses to the null floor, so the leverage gap becomes large.

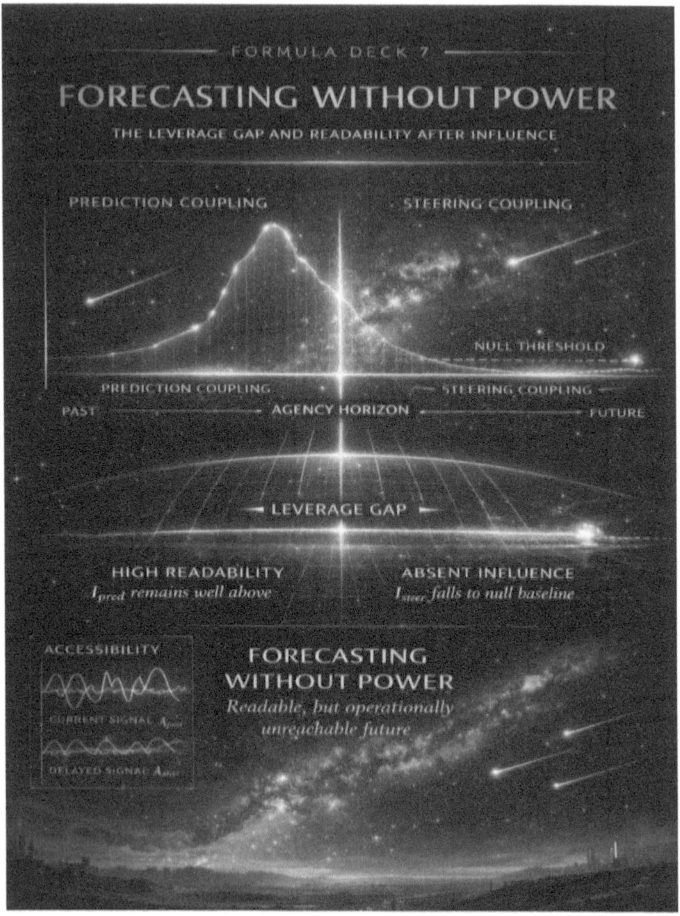
FORMULA DECK 7
FORECASTING WITHOUT POWER
THE LEVERAGE GAP AND READABILITY AFTER INFLUENCE
PREDICTION COUPLING
STEERING COUPLING
NULL THRESHOLD
PREDICTION COUPLING
STEERING COUPLING
PAST
AGENCY HORIZON
FUTURE
LEVERAGE GAP
HIGH READABILITY
I_{pred} remains well above
ABSENT INFLUENCE
I_{steer} falls to null baseline
ACCESSIBILITY
CURRENT SIGNAL
DELAYED SIGNAL
FORECASTING
WITHOUT POWER
Readable, but operationally
unreachable future

Chapter 8

Experimental Confirmation

If you want to claim that the future can stay readable after influence collapses, you cannot rely on clever storytelling. You need a test that can embarrass you. You need a setup where the world is allowed to say no, and where you cannot talk your way out of it with interpretation.

So the experiment begins with a simple promise. You build a little world with rules. Not a fantasy world, a repeatable one. The rules produce trajectories through time. The trajectories can drift, wobble, accelerate, settle, fail. You can run the same world again and again, with the same knobs, and watch what changes when you add delay.

Delay is the villain here, but it is also the gift. Delay is honest. It takes the drama out of arguments about will and replaces it with timing. When delay is small, steering is possible. When delay grows, steering begins to fade. The question is not whether steering fades, it does. The question is what happens to prediction at the same moment.

Most people assume prediction and steering rise and fall together. If you can predict, you can act. If you cannot act, you must not be able to predict. That assumption feels natural because in daily life prediction and control often appear together. You predict a curve, you turn the wheel. You predict a conversation will go badly, you choose your words. You predict a pot will boil over, you lower the heat.

But those examples live inside short loops. They are small worlds with fast feedback. Modern life often lives in long loops. Long loops break the intuition.

The measurement you want is therefore two sided. One measure asks how readable the future is from what you can see now. The other measure asks how much your actions actually change what happens later. These measures are cousins, but they are not twins. In a short loop they often travel together. In a long loop they can divorce.

To make this divorce visible you do something that feels almost too straightforward. You repeat the world many times and you vary only one thing, the delay between when you choose an action and when that action has a chance to matter. You do not allow yourself to change the rules to rescue your favorite conclusion. You keep the world fixed and you push it through different delays like a prism.

The procedure looks boring from the outside. You choose a delay. You run many trials. You collect outcomes. You compute readability. You compute steering. Then you change the delay and do it again. Then again. Then again. A long ladder of delays.

The boredom is a feature. It means the experiment is not relying on a single dramatic run. It is building a curve. It is mapping a boundary.

There is another trick that makes the test hard to cheat. You do not trust a single run. You do not trust a single seed. You do not trust one lucky configuration. You use many repeats and you look for a pattern that survives repetition.

If the effect is real, it will persist across runs. If it is a mirage, it will dissolve when you rerun it. A strong claim needs this kind of cruelty.

This is the emotional tone of the experiment. Not hope. Not gloom. A kind of scientific honesty that feels like a courtroom. You want a witness that cannot be bribed by your expectations.

The courtroom has one more rule. You must distinguish true influence from fake association. In time series, fake association is everywhere. The world is full of shared trends. If both your actions and the world state respond to the same background drift, you will see correlation even when your actions did nothing. If you are not careful, you will declare victory because the universe happened to move in a direction you expected.

So the design begins by admitting a humiliating truth. Correlation can lie. A clean test must include a way to estimate how much apparent steering could arise even if the steering were fake. That is the baseline. That is the null floor. That is the level you must beat before you are allowed to say influence exists.

When you build the experiment this way, you are no longer asking an abstract question like free will. You are asking a physical question. At what delay does the world stop answering you, even though it may still be readable.

That is the core of the confirmation program. It is not mystical. It is not rhetorical. It is a timing map.

Two.

Once you accept that correlation can lie, you need controls that kill the lie without destroying the boring structure of the world. The control must preserve the feel of the data while removing the one thing you want to test, real steering.

There are two simple control moves that accomplish this. They sound childish, which is why they are powerful.

The first is shuffle. You take the action choices and you scramble their pairing with outcomes, while keeping the overall distribution of actions the same. You still see the same range of actions, the same frequency of actions, the same typical magnitudes. But any genuine causal alignment between a specific action at a specific time and a later outcome is broken. If your steering measure stays high under shuffle, then your measure is not measuring steering. It is measuring a shared trend or an artifact.

The second is shift. You slide the action stream in time, so that the actions are preserved as a sequence but no longer line up with the outcomes they supposedly influenced. This is stricter than it sounds. It preserves the internal rhythm of action. It preserves the fact that actions are not random white noise. But it breaks the time alignment that would make steering real.

These two controls are the skeptical friends you bring to every exciting result. They keep you from believing your own story too early.

Now comes a subtle point that saves the whole enterprise from a common trap. Many measures of association spike at zero delay. They spike because of shared state, because the action and the current state can be coupled instantly in trivial ways, or because the measurement cannot separate what is simultaneous from what is causal. This spike is diagnostic. It tells you your measurement is sensitive to immediate alignment. It does not tell you that delayed influence exists.

So you treat zero delay like a bright reflection in a telescope. It is real, but it is not the planet you are looking for. You do not build your claim on it. You explicitly ignore it when discussing delayed influence.

Then you do something that makes the whole story operational. You do not declare influence just because a number is above zero. You declare influence only if it rises above the null floor generated by your controls. This is the move that prevents self deception.

In plain language it means this. If shuffled actions produce about the same steering score as real actions, then your steering score is not a steering score. If shifted actions produce about the same steering score as real actions, same conclusion. The world has exposed your instrument.

So you calibrate. You take the controls seriously enough that they define your thresholds. The null floor becomes part of the definition of influence.

There is also a second reason to love this calibration. It makes the experiment portable. It stops being a story about one choice of parameters. It becomes a method. Wherever you go, you bring the same skeptical friend, the same null controls, the same demand that influence must beat its own fake shadow.

At this point you have a machine for honesty. You can run it across delays and ask two questions at every delay.

Does the future remain readable.

Does action still matter beyond what the null controls would produce.

If readability stays high while steering drops to the null floor, you have the regime you were looking for. Readable future, no operational grip.

You can also do one more thing that makes the result feel less like a single cliff and more like a landscape. You can track failure rates. You can define a failure event, a boundary crossing that the world sometimes hits. Then you can see whether steering dies early while failure arrives later. That separation is emotionally important because it mirrors lived experience. You lose grip, then you watch the outcome proceed, sometimes for a long time, toward the boundary.

This is where the experiment stops being abstract. It starts to look like life. You can still see what is coming. You cannot change it. The cliff arrives on schedule.

The controls keep the story honest. They force the reader to trust the curve rather than your voice.

Three.

When you run this test across a wide range of delays, the picture that emerges is strikingly stable.

First, there is a small delay after which steering collapses to the null floor. In the confirmation suite that anchored the model, that post zero horizon appears at a delay of about four steps. That is not a magical number. It is an operational boundary under that world configuration. But the existence of a boundary is the point. The loop stops answering you beyond a modest delay, even though you are still acting.

Second, prediction does not collapse at the same moment. Readability stays substantial well beyond the steering horizon. In that same confirmation suite, prediction peaks later, around a delay near nine steps, and remains meaningfully above the null well into the region where steering is already indistinguishable from noise. Even near the later failure cliff, prediction remains present.

That is the heart of forecasting without power. The future stays legible after your grip dies.

Third, the failure event does not arrive at the steering horizon. The world can remain apparently stable for a long stretch while steerability is already gone. In the same confirmation suite, a near certain failure cliff appears much later, around a delay near thirty-five steps, long after steering has collapsed.

This temporal separation matters because it matches the lived shape of late clarity. You lose influence early, but the consequences play out later. The corridor forms. The ending becomes readable. You watch it approach.

The controls behave exactly the way you want if the effect is real. Under shuffle and shift, steering drops to near zero across delays beyond the immediate alignment region. The null floors remain tiny. That means the steering signal in the real data is not a spurious artifact of shared trends. It rises above its fake shadow only where true coupling exists.

The result is not just a pretty curve. It is a falsifiable pattern with teeth.

If someone claims the regime is an illusion, they have a clear way to attack it. They must show that the same separation appears under null controls. Or they must show that steering remains above null beyond the supposed horizon. Or they must show that prediction collapses whenever steering collapses. Or they must show that the horizon is a statistical mirage that vanishes across repeated runs.

But in repeated runs the pattern persists. The steering horizon is stable. The prediction ridge persists beyond it. The null controls remain low. The separation is not a one off accident. It is a regime.

Now translate that into human language without losing the precision.

The world can become easier to read because it is narrowing. The narrowing can be structural. It can be logistical. It can be built from delays. It can be built from limited energy. It can be built from crowded signals. Whatever the cause, the narrowing creates predictability.

At the same time, the same narrowing can destroy your ability to steer because action arrives late relative to the branch points that matter. Your motion becomes indistinguishable from background drift. You can still act, but the world no longer answers in time.

The experiment confirms that this separation is not only plausible, it is expected in delayed environments. It shows that a future can be readable long after it is not reachable.

This is the moment where the reader should feel a shift. Not despair. Relief of understanding. The nausea of late clarity becomes less personal. It becomes a measurement. It becomes something you can point to and say, the lever was no longer coupled, the window was gone, the action channel had collapsed, even though the observation channel remained bright.

And that is the practical value of confirmation. It turns an emotional experience into a design target.

If you want less forecasting without power, you do not merely improve prediction. You protect steerability. You move levers upstream. You shorten loops. You preserve reversibility. You reduce overload so meaning arrives on time. You respect energy because energy is part of timing. You demand that warnings come with reachable actions inside usable windows.

The experiment does not tell you how to build a perfect world. It tells you what not to confuse.

A clean forecast is not proof of fairness.

A clean forecast is not proof of control.

A clean forecast can coexist with dead levers.

Once that is proven in a repeatable way, you can stop arguing about feelings and start arguing about windows.

The split between readability and steerability is not philosophical, it is mechanical, and once you see the mechanics you stop being surprised by late clarity.

Prediction is about patterns in what you can observe, steering is about changing what happens next. They can travel together in short loops, they can separate in long loops. That separation is the whole point.

Readability can improve for reasons that have nothing to do with your levers. You can add sensors. You can add logs. You can add resolution. You can add memory. You can add models trained on huge collections of past cases. All of that can make the future easier to infer from the present. The map gets sharper.

Steerability depends on something more fragile. It depends on timing. It depends on whether an action applied now produces a distinct change later that rises above background drift, noise, and interference. It depends on whether the world answers your motion soon enough that you can correct again before the branch closes.

That is why a world can become more predictable while becoming less steerable. The observational beam improves. The action loop stays long, costly, delayed, and crowded.

A second reason the separation becomes common is that many processes become more constrained as time passes. Early on there are many viable branches. Later on branches collapse into corridors. As corridors form the outcome becomes easier to predict because fewer exits remain. Readability rises because the future has fewer shapes remaining. At the same time steerability falls because the same narrowing removes options.

That is why late clarity feels sharp. The mind senses a corridor before it names it. The corridor has an emotional texture, a sense that the decision has already been made somewhere. Often the corridor is made of accumulated delays. Signals were present. Levers were earlier. Windows were small. By the time the corridor is obvious the only remaining actions are large and blunt.

Large and blunt actions do not restore steerability in a delayed loop. They often create overshoot. Overshoot creates side effects. Side effects create noise. Noise slows meaning. Slower meaning shortens the remaining window. So sincere action can still be punished by timing when it lands too late.

This is a harsh lesson. The later you are, the more force you feel you need. The more force you apply, the more you destabilize the thing you are trying to correct. Late force becomes a generator of volatility.

Now add a third ingredient that is easy to underestimate. Measurement itself can seduce you into motion that is not control. When you can see more, you feel compelled to react more. When you react more with delayed feedback you create oscillation. You create noise with your own hands. The loop becomes jumpier because you are chasing.

Chasing is what happens when you aim your actions at a world that is already gone. You see a number rise. You react. The reaction lands after the number has already changed. You react again. Many panics are chasing. Many arguments are chasing. Many collapses are chasing. A delayed loop trains you to chase because your measurement arrives after the state moved.

This is why more measurement does not guarantee more control. Seeing and steering are different couplings. You can build a forecast that is nearly perfect. If the lever is not reachable inside the effective window the forecast becomes a beautiful report.

A modern environment often improves the viewing channel far faster than the action channel. It becomes very good at telling you what is happening. It remains slow at letting you change what happens. So you get a familiar experience. A person receives fine grained knowledge of risk. Then the person is routed into slow channels for action. They are told to wait. They are told to submit. They are told to confirm. They are thanked. Meanwhile the outcome proceeds because the branch point does not pause.

Another modern delay is overload. Attention is finite. When signals multiply detection time grows. A warning can arrive early on paper and arrive late in the mind. The brain must notice it. The brain must classify it. The brain must believe it. The brain must translate it into a plan. The body must execute the plan. Each step consumes time and energy. When the environment is flooded the mind delays.

This creates a subtle split. An environment can claim the warning arrived early. A person can experience it as late. Both can be true. A flood of prompts can destroy timing without touching a clock.

Some environments respond by adding even more prediction. It sounds sensible. Warn earlier. Give more time. But if the lever remains delayed, earlier warning does not help. It can harm. It extends the duration of helpless awareness. It turns foreknowledge into stress. Stress narrows attention. Narrow attention delays meaning. Delayed meaning shrinks the usable window. Then more alerts are issued. The cycle tightens.

That is one reason a culture can become highly foreseeing and highly anxious at the same time. It sees more. It can do less. It feels responsible for what it cannot reach.

Now walk through a few domains where the mechanics are easy to feel.

In medicine prediction can improve dramatically. Trends can be detected early. Risk can be scored. A case can be flagged before the visible crisis. But steerability depends on staff, time, beds, authority to act, and the ability to intervene while the body is still reversible. If the warning arrives but the intervention is delayed by overload and friction then prediction becomes a report about an approaching cliff. It can even become a moral weapon, you were warned, you should have acted, while the lever may not have been reachable.

In finance markets can price in information rapidly. That can make some outcomes more predictable. At the same time individual actors often have levers that are slow. Credit decisions and liquidity access and emergency buffers do not appear on demand. So risk becomes visible early while the ability to adjust position is limited by delay and cost. People can watch the future tighten, then be forced to move only when the corridor is already narrow and the price of motion is highest.

In supply chains sensors can show disruption earlier. But steering requires inventory, alternate routes, coordination across firms, and time margins. When everything is tuned for minimal slack steerability collapses. The environment becomes predictable because once one constraint binds many downstream outcomes become almost certain. Prediction rises. Power falls.

In relationships the instruments are not sensors and models, they are memory and pattern recognition. A person can forecast the ending because the signs are there. But steering requires short loops. It requires early repair. It requires timely honesty. If those levers are delayed the corridor forms. The ending becomes predictable. At the same time it becomes hard to change because by the time it is obvious the internal state has hardened.

Across these domains the geometry repeats. A readable future does not imply a steerable future.

There is also a deeper reason prediction can remain strong after leverage collapses. Prediction can piggyback on structure that is not controllable by the observer. If a process is dominated by large background forces then an observer can infer its direction without being able to alter it. Weather is the clean example. A person can know the storm track. A person cannot move the storm. A person can only move themselves, and even that depends on levers that might be delayed. A useful diagnostic follows. When late clarity hits, ask whether the levers you needed were local, early, and small, and whether they were reachable when the first meaningful signals appeared. If they were not reachable then the experience is not personal failure, it is coupling failure.

A second diagnostic follows. If an environment gives ever more detailed forecasts but does not shorten the time between action and consequence then it is investing in readability while neglecting steerability. The result is a culture of spectators, highly informed, constantly reacting, rarely able to alter trajectory in time.

The most important practical rule is blunt. Prediction is valuable only when it arrives attached to timely levers inside usable windows, with reversibility still available, with signals clean enough to be detected without heroic attention. Otherwise prediction is not guidance. It is exposure.

Exposure feels like responsibility. It feels like you have been handed the future so you must be able to change it. That feeling is an illusion created by information. The world does not change because you were shown it. The world changes only when an action couples to outcomes in time.

The result you are looking for is not a dramatic single moment, it is a landscape that keeps its shape when you change the random seed, when you repeat the run, when you shuffle the order, when you try to fool yourself.

When you sweep delay step by step, two curves appear that refuse to stay married.

One curve describes how readable the future remains from what you can observe now. The other curve describes how much your chosen actions still change what happens later in a way that is distinguishable from background drift.

In short loops these two curves tend to rise and fall together, which is why everyday intuition treats prediction and control as the same thing. In long loops they separate, and the separation is not subtle.

As delay increases, the influence curve drops first. There is a clear point where actions stop leaving a signature that rises above the null floor defined by the controls. Beyond that point you are still acting, you are still choosing, you can still press every button you can find, but the world no longer answers in a way that can be measured as steering.

That point is the agency horizon in operational form. It is not a philosophical boundary. It is where the data stop supporting the claim that action has leverage beyond what the controls can fake.

In the confirmation runs that anchor this model, that horizon appears quickly. After a modest delay, on the order of a handful of steps, steering drops to the same tiny level produced by shuffle and shift controls. The exact number is not sacred. What matters is the structure. Influence collapses early.

The prediction curve does not collapse at the same time. It stays high beyond the horizon. In many runs it even peaks after influence is already gone. The future remains readable, sometimes remarkably readable, even when the action channel is dead by the same standard that the controls enforce.

This is the central confirmation. It is not a story. It is what the curves show once the test is designed so that false steering is punished.

Then there is a third curve that makes the entire picture feel uncomfortably familiar. Failure.

You define a boundary crossing, a collapse event, a threshold that marks a world that has drifted into a bad region. You can call it a failure cliff because that is how it feels when you look at the statistics. The probability of failure stays low for a while, then it rises sharply, then it becomes nearly certain.

Here is the striking feature. The failure cliff appears much later than the steering horizon.

Influence dies early. The bad outcome arrives later.

There is often a long stretch where the world looks stable enough, where the trend is readable, where warning signs are visible, yet where influence has already collapsed below threshold. You are watching the corridor form. You can describe it. You can predict it. You cannot steer it back, not in a way that beats the null floor.

This is why the confirmation matters for ordinary human experience. It explains a very specific kind of suffering. You lose leverage and then you keep living for a while. You keep receiving updates. You keep seeing the ending become clearer. The clarity is real, but the grip is gone. You are not being punished by fate, you are living inside the separation between observation and influence.

The controls are what make the story trustworthy.

Shuffle destroys time alignment while preserving the overall distribution of action. Shift preserves the rhythm of action but breaks its alignment with outcomes. In both cases, any steering score that survives is not steering, it is shared structure, a background trend, an artifact of measurement. Those controls generate a null floor.

When real steering falls to that floor, you have an operational reason to stop calling it steering.

This is also why the spike at zero delay must be treated as diagnostic only. Zero delay can contain trivial alignment that looks like influence without being influence across time. Many measures will light up at zero delay because the world state and the action choice share information instantly. That is not the claim. The claim is about delayed influence, the kind of influence that matters when real levers are slow.

So you ignore zero delay when making the important statements. You care about the post zero region because that is where daily life lives, where people wait, where meaning arrives late, where action takes time to land.

Now step back and notice what has been confirmed.

A delayed world can keep the future readable after leverage collapses. This is not rare. This is not a corner case. This is an attractor regime of long loops with limited energy, noisy measurement, delayed action, and narrowing corridors.

Once you accept that, a number of modern social habits look different.

It becomes clear why more forecasting can produce more anxiety without producing more control. More forecasting extends the duration of helpless awareness. Helpless awareness is stress. Stress narrows attention. Narrow attention increases internal delay. Increased internal delay shrinks effective windows further. The loop lengthens from the inside.

It becomes clear why constant monitoring can feel like action and still fail to produce change. Monitoring improves readability. It does not automatically improve steerability. In a delayed world, monitoring can even make you worse at steering because it burns energy and pushes you toward chasing, reacting to fluctuations that your delayed actions cannot stabilize.

It becomes clear why late warnings can be socially toxic. A late warning does not open options. It produces receipts. It produces blame. It produces arguments about responsibility that ignore the timing of levers.

This is where the confirmation turns into a design rule.

If you want fairness, you must protect steerability, not celebrate visibility.

If you want safety, you must keep effective windows positive in lived time, not merely in policy language.

If you want dignity, you must refuse to confuse being informed with being empowered.

The experiment also gives you a way to evaluate claims.

Any environment that improves prediction while leaving action delayed is drifting toward forecasting without power.

Any environment that produces better dashboards while shrinking undo windows is drifting toward forecasting without power.

Any environment that floods attention with alerts while hiding the few levers that actually matter is drifting toward forecasting without power.

And any environment that points to warning delivery as proof of fairness, while the effective window is already closed, is using prediction as anesthesia.

The antidote is not mystical. It is timing craftsmanship.

Shorten loops. Move levers upstream. Keep actions reversible long enough to be usable. Reduce clutter so detection happens fast. Respect human energy because energy is part of timing. Make the moment of branch closure visible so blame cannot float free of physics.

You can also see why the most meaningful improvements will often look boring.

A longer undo period is boring. It is also steerability.

A clearer warning that is rare and specific is boring. It is also signal integrity.

A policy that places levers earlier in the causal chain is boring. It is also freedom that can be measured.

A design that reduces the number of steps between intent and effect is boring. It is also coupling.

This is the deeper reason the confirmation program matters. It does not merely say that late clarity exists. Everyone already knows that emotionally. It says late clarity is measurable, it has a boundary, it has controls that can expose fake signal, and it has remedies that are not moral sermons but engineering choices.

A world that becomes more readable does not automatically become more steerable. That is the lesson. It is a lesson that can be proven in a repeatable test, and that is why it belongs in science rather than in mood.

Once you see the curves separate, you stop treating your own late clarity as a personal verdict. You start treating it as a map.

The map tells you where leverage died.

The map tells you that prediction can remain strong anyway.

And the map tells you the only honest response, protect the parts of life where the world still answers in time.

Forecasting Without Presence
(FAWP)
Seeing the unsteerable is not the same
as steering the unseeable.
STEERING
PREDICTION
NULL FLOOR
Prediction persists
beyond steering collapse
Readable but unreachable
futures emerge
NULL FLOOR

FORMULA DECK (Chapter 8)
Experimental Confirmation, Null Floors, and the Verified FAWP Regime

1. Goal of the confirmation program. Consider a repeatable world generating trajectories over time. The aim is to measure two distinct couplings as a function of horizon τ. First, readability of the future from current observations. Second, influence of current actions on future outcomes. The confirmation claim is that readability can persist after influence collapses, producing a verified Forecasting Without Power regime.
2. Generative process and repeated trials. Let the world produce sequences of states, actions, and observations over time. Over many independent trials, the same generative process is run with identical parameter settings. Delays and noise are included. The key requirement is repeatability, meaning the same configuration produces comparable ensembles across runs.
3. Horizon sweep and measurement ladder. Fix a grid of horizons τ ranging from zero up to a maximum delay. For each horizon, measure prediction coupling and steering coupling using the trial ensemble. The outcome of the confirmation suite is a pair of curves over τ, one describing readability of the future and the other describing influence of action, together with null

control curves used for calibration.

4. Prediction coupling estimator. Prediction coupling at horizon τ measures how strongly future states are statistically related to current observations. Operationally, it is estimated from the empirical joint distribution of present observations and future states across trials and times. Estimation may be pooled across trials or stratified by state bins to reduce mixing artifacts.
5. Steering coupling estimator. Steering coupling at horizon τ measures whether chosen actions measurably influence future states beyond what current observations already imply. Operationally, it is estimated by testing whether actions and future outcomes remain coupled after accounting for accessible information at decision time, using the same pooling or stratification scheme as prediction.
6. Zero horizon diagnostic and exclusion rule. The value at $\tau = 0$ can exhibit a large diagnostic spike due to instantaneous shared state or estimator artifacts. The confirmation program treats $\tau = 0$ as diagnostic only and excludes it from delayed influence claims. All causality relevant statements are made for nonzero horizons.
7. Shuffle null control. Construct a shuffled action control by permuting the action sequence across trials or time indices while preserving the marginal distribution of actions. Compute a null steering curve using the shuffled actions paired

with original future states. This breaks causal alignment while preserving action statistics.

8. Shift null control with autocorrelation preservation. Construct a time shift control by shifting the action stream by a lag chosen to break alignment with future states while preserving action autocorrelation. Compute a second null steering curve using shifted actions paired with original future states. This preserves action rhythm while breaking temporal alignment.
9. Null floor and conservative thresholding. Define a null floor distribution for steering coupling from the shuffle and shift controls. Choose a conservative threshold from the upper tail of this null distribution, for example the 0.99 quantile. Declare steering present at a given horizon only when measured steering coupling exceeds this threshold. This prevents false leverage claims driven by spurious correlation.
10. Extraction of the post zero agency horizon. Define the post zero agency horizon as the smallest nonzero horizon at which steering coupling falls to the calibrated null floor. Beyond this horizon, influence is operationally indistinguishable from null even though actions continue to be taken.
11. Verified FAWP regime. The FAWP regime is verified when prediction coupling remains significantly above noise at horizons where steering coupling has collapsed to the null floor.

This confirms the existence of readable but unreachable futures under delayed and noisy access.

12. Confirmation robustness requirements. The confirmation suite should be repeated across multiple seeds and parameter settings. A verified FAWP claim requires that the same qualitative pattern persists, a $\tau = 0$ spike treated as diagnostic only, a nonzero agency horizon identified by null thresholding, and a readable region beyond that horizon where prediction remains strong while steering remains at the null floor.
13. Optional early warning signature. In some unstable regimes, prediction coupling can rise sharply as control collapses. This is an early warning signature, not a proof of influence. It is treated as a deployable signal only when it remains stable under null controls and persists across repeated runs.

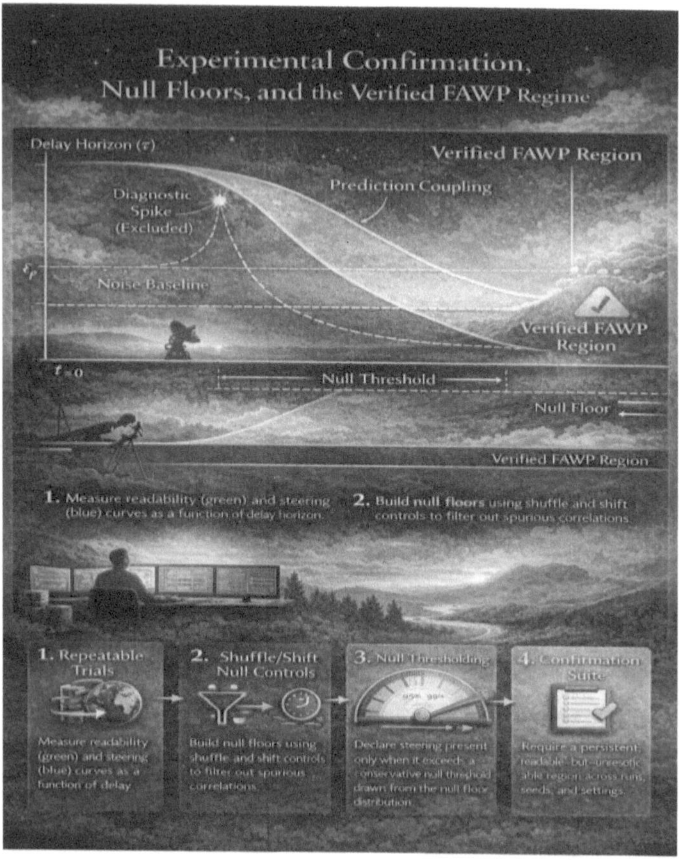
Experimental Confirmation,
Null Floors, and the Verified FAWP Regime
Delay Horizon (τ)
Verified FAWP Region
Diagnostic
Spike
(Excluded)
Prediction Coupling
Noise Baseline
Verified FAWP
Region
t = 0
Null Threshold
Null Floor
Verified FAWP Region
1. Measure readability (green) and steering (blue) curves as a function of delay horizon.
2. Build null floors using shuffle and shift controls to filter out spurious correlations.
1. Repeatable Trials
Measure readability (green) and steering (blue) curves as a function of delay
2. Shuffle/Shift Null Controls
Build null floors using shuffle and shift controls to filter out spurious correlations.
3. Null Thresholding
Declare steering present only when it exceeds a conservative null threshold drawn from the null floor distribution.
4. Confirmation Suite
Require a persistent,
seeds, and settings.

Chapter 9

Detectors and Deployment

The modern world has learned to warn you. It has not learned to give you handles.

You can be told that a storm is coming, that a patient is trending toward collapse, that a supply chain is about to fracture, that a market is about to seize, that a relationship is quietly bleeding out. You can be told early. You can be told with confidence. You can be told in graphs and colors and polite sentences. And still you can be too late in the only sense that matters, too late to change the outcome.

This is why early warning has become a strange kind of cruelty. Not because warning is bad, warning can be beautiful. Because warning is often delivered as proof of fairness, when it is only proof of visibility. Visibility is not power. A red light is not a lever. A prediction is not a handle. Information can be a gift, and information can also be a receipt.

A real detector is not a device that predicts well. The world is already full of those. A real detector is a device that tells you whether the future is still steerable from where you are standing, with the levers you can actually reach, in the time you still have.

That sounds like a small distinction. It is enormous.

Prediction answers, what is likely to happen. Detection answers, can I still change what happens in a measurable way. Those two answers can diverge, and when they diverge, most people keep reacting as if they are the same.

That reaction is natural and it is also expensive. It creates the most common modern failure mode, obsessive monitoring paired with late action. You stare at the forecast harder, you refresh the dashboard more often, you consume more updates, and the world continues to slide into the outcome anyway. You did not lack knowledge. You lacked coupling. You lacked a short enough loop.

A detector that matters must therefore be built on a harsher foundation than comfort. It must be built on the idea that you do not get to call something influence just because it correlates. Correlation is cheap. Correlation is everywhere. Correlation will flatter you and then abandon you. A good detector assumes that the world can produce associations that look meaningful even when your action did nothing. A good detector demands a baseline of fake steering and refuses to be impressed by anything below that floor.

In plain language, it does not ask whether your action and the outcome moved together. It asks whether your action changed the future more than it would have changed under a world where your action was only background noise.

This is why a detector has to be calibrated like a courtroom, not like a motivational poster. The courtroom does not care how sincere you feel. It cares about what survives cross examination.

The first cross examination is simple. If you scramble your actions and the result stays the same, then your so called influence was never influence. It was a shared drift. It was a trend. It was a coincidence produced by structure.

A detector that cannot be embarrassed by this test is not a detector, it is a storyteller.

The second cross examination is equally simple. If you slide your actions in time, keeping the rhythm but breaking the alignment, and the result stays the same, then your influence was not influence. It was a background pattern that your measure cannot separate from causation.

These tests create something precious, a null floor. The null floor is what your world produces when steering is fake. It is the level of apparent influence that can exist even when you have no real grip.

Most public warnings ignore this. They point to a number above zero and call it control. They point to a chart and call it agency. They do not measure their own fake shadow, so they cannot tell the difference between a lever and a decoration.

A detector that matters does one thing first. It learns how easy it is to fool itself.

Once you have a null floor, the problem becomes clean. You look for the region where prediction remains strong but steering has fallen to the null floor. That region is the dangerous one, because it produces the strongest illusion. The future looks readable, so you feel as if you should be able to do something. But your action channel is already too late, too weak, too delayed, too drowned, to change the future in a measurable way.

This region is where people suffer. It is also where institutions love to hide. Because the region is perfect for transferring blame. If you can predict, you must be responsible. If you were informed, you should have acted. If you did not act, you failed.

A detector ruins that trick. It separates being shown from being able. It separates warning from handle. It tells you whether the lever is real.

There is a second reason this region matters. It can appear before the obvious collapse. The world can remain stable looking for a long time while steerability is already gone. The corridor forms quietly. The ending becomes legible. The cliff arrives later. The detector that matters is the one that tells you that the lever died early, while there is still time to move levers upstream, widen reversibility, shorten loops, reduce noise, protect energy, clean the signal.

In other words, it gives you a chance to change the design, not just brace for the outcome.

This is the difference between a warning that sedates and a warning that mobilizes. A sedating warning gives you a forecast and no handle. It increases helpless awareness. It increases anxiety. It burns attention. Burned attention slows detection. Slower detection shrinks the usable window further. The warning becomes part of the collapse. It becomes another delay.

A mobilizing warning is attached to an action that is reachable now, not later. It is attached to a lever that still couples to outcomes above the null floor. It is attached to reversibility, so early action is safe. It is delivered with clean signal integrity, so it is detected quickly. It respects human energy, so the mind can actually respond.

That is the standard. Anything below it is theater.

A society that understands this begins to treat alerts differently. It stops measuring success by delivery of information and starts measuring success by preservation of steerability. It stops congratulating itself for forecasting and starts auditing whether the levers were reachable in time. It stops treating a notification as protection and starts treating it as a claim that can be tested.

A detector is therefore not only a technical object. It is a moral object, because it changes what can be blamed. It makes timing visible. It makes coupling testable. It turns vague arguments about responsibility into an audit of access.

This matters because responsibility without reachable access is not responsibility, it is superstition with better vocabulary. People cannot be judged by what they could not reach. They cannot be made accountable for outcomes that were already beyond their operational horizon. They can be held to standards only when the world gave them a real chance to change what happened.

That is why the most important output of a detector is not a dramatic alert. The most important output is a timing receipt. A simple record of when the signal became available, when the lever was reachable, when the branch closed, and whether the lever had real coupling above the null floor.

The receipt is the antidote to hindsight. Hindsight is the great distortion. Hindsight makes everything look obvious. Hindsight makes every corridor look like a choice. Hindsight makes every late warning look like an opportunity.

A timing receipt restores reality. It says, this is when meaning could have arrived. This is when action could have mattered. This is when the door closed.

Once you have that, you can stop arguing about feelings and start improving the loop. You can move levers upstream. You can widen undo windows. You can clean signals. You can remove friction from urgent action. You can reduce overload so detection is not delayed inside the mind. You can protect energy so the human controller remains fast enough to respond.

This is what it means to deploy prediction ethically. It means refusing to deliver forecasts as anesthesia. It means refusing to confuse visibility with fairness. It means insisting that every warning carries a handle, or else it is labeled honestly as a receipt.

The future will keep becoming more readable. That is almost guaranteed. The harder achievement is to keep the world steerable where steerability still matters. A detector is the tool that tells you when you are losing that achievement, early enough to do something real, not heroic, real, the boring work of shortening loops and preserving windows before they close.

A detector that matters has to be built like an instrument, not like a mood. It has to give you the same answer when you are calm and when you are panicking. It has to resist flattery. It has to resist coincidence. It has to refuse to confuse shared trends with real leverage.

So you begin by measuring two different things that people keep mixing in their heads.

The first is readability. How much the present contains information about what comes later. You can estimate this from ordinary observation streams, signals, logs, measurements, even human judgments if they are consistent. The details vary by domain, but the idea stays the same. When readability is high, the future is legible from what you can see now.

The second is steerability. Whether actions you take now actually change what happens later in a way that rises above background drift. This is the harder one, because the world is full of patterns that look like influence even when influence is absent. In a crowded environment, the outcome can move and your actions can move and the two can seem connected because both are being pushed by the same hidden forces.

That is why a detector needs a null floor.

The null floor is the level of apparent influence you get when steering is fake. It is not a philosophical baseline, it is an empirical one. You generate it by breaking alignment between action and outcome while keeping the boring structure intact.

You do this in two simple ways.

You scramble the pairing between actions and outcomes while keeping the action distribution the same. This preserves how often actions occur and how large they are, but it destroys the idea that a particular action at a particular time caused a later change.

You also slide the action stream in time, so the rhythm is preserved but the alignment is broken. This keeps the fact that actions have momentum and habit and bursts, but removes their supposed timing relationship to outcomes.

These two moves tell you how easy it is for your influence measure to hallucinate. They create the shadow version of steering. That shadow is what any honest detector must beat.

Now you set a rule that sounds almost too strict, but it is the whole point. You do not call something leverage unless it rises above the null floor by a margin that is hard to achieve by accident.

In practice this means you choose a conservative threshold, the kind that keeps false positives rare. You look at the distribution of fake steering under the null controls, and you place the threshold high enough that only a small fraction of null cases cross it. If the null almost never crosses, then crossing becomes meaningful. If the null crosses all the time, then your detector is a siren, not an instrument.

This is also why you cannot build a detector from a single number in a single moment. A single spike is cheap. Noise produces spikes. Luck produces spikes. Even genuine influence can produce spikes that do not last. What you need is persistence.

Persistence is the requirement that the pattern holds for several consecutive windows. The exact count depends on the domain, but the philosophy is universal. One moment of apparent leverage can be coincidence. Several moments in a row, each beating the null floor, is much harder to fake. So the detector uses three ingredients.

A readability signal. It says the future is legible.

A steerability signal. It says action still couples to the future.

A null calibrated threshold. It says whether the steerability signal is real or just the shadow.

From these ingredients you can build a practical object that people can actually use. You track a window of time, or a ladder of delays, and you ask a simple question at each step.

Is readability high.

Is steerability above the null threshold.

If readability is low, you are in fog. The detector does not pretend otherwise. In fog, the honest output is uncertainty, and the correct response is often to shorten loops and improve signal integrity, not to demand prophecy.

If readability is high and steerability is high, you are in the healthy region. The future is legible and you still have a grip. This is where small early actions are most effective, because the loop still answers.

If readability is high and steerability is at the null floor, you are in the dangerous region. The future is legible but the grip is gone. This is where forecasts turn into receipts. This is where blame games thrive. This is where people develop late clarity and call it personal failure.

That region is the operational detection window. It is the interval where you can still see the cliff forming, but direct steering no longer works. It is the moment where the correct move is not to push harder on the same levers, but to change the architecture. Move levers upstream. Increase reversibility. Reduce friction. Clean signals. Protect energy.

Shorten loops. If you do not do that, you will be reduced to watching.

A detector becomes truly valuable when it has lead time. It must warn before the failure accelerates. If it only triggers when the outcome is already inevitable, then it is not a detector, it is a memorial.

So you test lead time the same way you tested steering, with skepticism. You compare when the detector would have triggered to when failure rates begin to rise sharply. You want the trigger to come earlier, consistently, across repeated runs. If it does, then the detector is not just describing the corridor, it is giving you time to widen it.

This is also where the calibration work pays off. False positives are not a minor annoyance. False positives destroy trust. They teach people to ignore alerts. They burn attention. Burned attention creates internal delay. Internal delay shrinks effective windows. A noisy detector can make the world less steerable by exhausting the human controller.

So you enforce a discipline. You keep the false positive rate low under null controls. You require persistence. You avoid triggering on the zero delay region where trivial alignment can fool measures. You insist on repeated evidence that survives the shadow tests.

When you build the detector this way, it stops being a dramatic alarm and becomes a quiet instrument. It does not shout. It does not moralize. It does not pretend to be certain in fog. It simply answers the only question that matters.

Is the future still coupled to your actions in time.

If yes, act early and gently.

If no, stop pretending that more monitoring will restore control. Use the warning as a design signal. The lever needs to be moved. The undo window needs to be widened. The friction needs to be removed. The signal needs to be cleaned. The human energy budget needs to be respected.

There is a small but important twist here. A detector is not only for individuals. It is also for environments. It can be used as an audit tool.

If an environment produces high readability and low steerability for long stretches, it is creating forecasting without power. It may be doing it accidentally through delay and overload. It may be doing it intentionally by placing levers behind friction. Either way, the detector makes the pattern visible. It makes it measurable. It prevents the environment from claiming that warning delivery equals fairness.

This is why the most useful output of the detector is often not a single alert. It is a record of timing. When readability rose. When steerability fell. When the corridor began. When failure accelerated. How long the operational window lasted. How often the detector cried wolf under null controls. How often it provided true lead time.

That record is what lets you improve the loop, rather than argue about who should have known. A good detector does not exist to impress you with intelligence. It exists to protect coupling. It exists to tell you, with calibrated humility, when you still have a grip, and when you are already being offered knowledge as a substitute for a handle.

A detector becomes real when it changes what you do.

If it only changes how you feel, it is another kind of content. If it only changes what you watch, it is another kind of monitoring. If it only changes the language of blame, it is another kind of moral theater. A real detector must reliably do one thing, it must create earlier reachable action in the domains where earlier reachable action is still possible.

That is why the last step is not more cleverness. The last step is deployment discipline.

The first deployment rule is brutal and simple. Never deploy a detector without attached levers.

If you tell a person the future is readable and closing, and you do not give them a handle that still couples, you are not helping. You are extending helpless awareness. Helpless awareness burns attention. Burned attention increases internal delay. Increased delay shrinks windows further. Your detector becomes part of the collapse.

So every alert must ship with an action that is reachable now, not later, not after a queue, not after ten confirmations, not after a negotiation. Reachable now.

And when stakes are high, the action must be reversible, because reversibility is how you make early motion safe. Without reversibility people hesitate. Hesitation is delay. Delay is the enemy.

This is why undo windows are not a luxury. They are the most efficient way to preserve agency under delay. An undo window converts fear into motion. It converts paralysis into early correction. It keeps the corridor wider.

The second deployment rule is about honesty in language. A detector must label what it is doing.

If it is in fog, it must say fog. If readability is low, the correct output is not prophecy, it is uncertainty. Uncertainty is not weakness. It is a precise statement that the observation channel is degraded. In fog, the correct remedy is usually to clean signal, reduce clutter, shorten loops, and protect energy, not to demand that a model produce a confident number.

If it is in a steerable region, it must say steerable. And it must encourage early gentle action, not late dramatic action. Early gentle action is how you keep the loop inside the stable region. Late dramatic action is how you create oscillation.

If it is in the readable but unreachable corridor, it must say it plainly. Not in a sentimental way, in a timing way. The lever is dead at this horizon. Action is indistinguishable from null beyond this point. The warning is no longer about steering, it is about redesign.

This is where most deployments fail, because environments love to pretend that seeing equals controlling. They love to give forecasts and call them empowerment. A detector that is honest forces a different posture. It says, you can see, and you cannot steer, so stop pretending the correct move is more monitoring. The correct move is structural.

The third deployment rule is signal integrity. A detector must be rare enough to be believed.

If it triggers constantly, it trains people to ignore it. If it triggers on small fluctuations, it becomes noise. If it triggers without persistence, it becomes panic. Panic increases delay. Delay shrinks windows. Again, the detector becomes part of the collapse.

So the detector must enforce calm. It must require sustained evidence. It must be conservative about declaring leverage. It must maintain a low false positive rate under the null controls, because false positives are not free, they are attention debt.

This is why calibration is not an academic detail. Calibration is how you protect the human controller. When you keep false positives low, you keep trust high. When you keep trust high, responses are faster. When responses are faster, the loop is shorter. When the loop is shorter, agency survives longer.

The fourth deployment rule is timing receipts.

A detector should not only emit alerts. It should emit receipts. Minimal records of timing, without content. When the signal became available. When the lever was reachable. When steerability dropped to the null floor. When failure rates began to accelerate. How much lead time existed. How many times the detector would have cried wolf under controls.

Receipts do two things at once. They enable improvement and they block superstition.

They enable improvement because you can look at the timing and see where the loop is failing. Is the lever arriving late. Is the warning buried. Is the action channel gated.

Is the undo window too short. Is overload delaying recognition. You can fix what the receipt reveals.

They block superstition because they remove the power of hindsight. Hindsight makes every ending look obvious. Hindsight makes every corridor look like a choice. Receipts restore the physics. They show where the effective window was positive, where it was zero, where it was negative. They show when the door closed.

Once you have receipts, responsibility becomes calibratable rather than emotional. The person is not judged by what they could not reach. The environment cannot hide behind, we warned you, if the lever was unreachable inside the window.

This is the fairness rule in operational form. Responsibility tracks reachable access, not hindsight.

Now you can see why a detector is also an ethical instrument. It changes what can be blamed. It shifts accountability upstream to whoever controls windows, levers, friction, and signal integrity. It makes it impossible to claim that warning delivery alone equals fairness.

And this leads to the final deployment rule, do not use prediction to sedate.

A sedating deployment floods people with forecasts, probabilities, and bright charts, then leaves them with little or no reachable action. It produces a culture of spectators and a culture of guilt. It teaches people to monitor instead of to steer. It treats visibility as virtue. It treats inevitability as maturity.

A mobilizing deployment does the opposite. It uses prediction as a trigger to move levers upstream. It uses prediction to widen undo windows. It uses prediction to shorten loops. It uses prediction to reduce friction when time matters. It uses prediction to protect energy by simplifying decisions, not multiplying them.

You can state the whole philosophy in one sentence.

If the warning does not widen the reachable set of futures, it is not help.

A detector is useful because it can tell you, in a calibrated way, when the reachable set is shrinking. But usefulness depends on what you do with that knowledge. If you respond by turning it into more content, you worsen the condition. If you respond by changing the architecture of action, you improve the condition.

So the endpoint is not an alert. The endpoint is a world in which alerts are less necessary because loops are short enough, levers are early enough, reversibility is wide enough, signals are clean enough, and human energy is respected enough that steerability survives.

That is what deployment means here. Not installing a tool, but protecting coupling.

Because in a world that keeps becoming more readable, the only dignity that remains is not knowing, it is being able to act while action still matters.

FORMULA DECK (Chapter 9)
Detectors and Deployment, Operational Windows, and Calibrated Alerts

1. Detector objective. The detector answers a practical question at time t. Is the future still steerable from the observer current access, or has steering collapsed to a null floor even though prediction remains strong. The detector is designed to separate genuine leverage from spurious correlation. The detector does not claim certainty. It produces calibrated outputs designed to be honest about what is and is not reachable.
2. Access and decision constraint. Let D_t denote the data accessible to the observer up to time t, and let F_t = σ(D_t) denote the corresponding information set. Actions are chosen under an access constrained policy, meaning actions are functions of what is actually accessible at time t.
3. Prediction coupling as readability input. For a nonzero horizon τ, prediction coupling measures how readable the future is from what the observer can currently see. High prediction coupling at delay τ means the future state is statistically predictable from present access. Prediction coupling can be high even when it reflects shared state or common causes. Readability alone does not imply leverage.
4. Steering coupling as leverage input. For a nonzero horizon τ, steering coupling measures

whether chosen actions measurably change future outcomes at delay τ beyond what access alone already implies. Actions include deliberate inaction. The policy defines what the observer actually does given current access. If steering coupling falls to a null floor, actions may still move an interface but they do not move the future in an operational sense.

5. Null controls and calibration set. Define a calibration set of null controls that break action outcome alignment while preserving basic structure. Examples include shuffled actions and time shifted actions. For each horizon τ, compute a null baseline for steering coupling using these controls.
6. Null quantile threshold. Choose a conservative quantile level β near 1. For each horizon τ, define a steering threshold as the β quantile of the null baseline at that horizon. Declare steering present at horizon τ only when measured steering coupling exceeds this threshold. This controls false leverage claims under null conditions. Parameter choices are domain specific. Higher β and stronger persistence reduce false positives at the cost of later alerts.
7. Post zero horizon gate. Exclude $\tau = 0$ from leverage claims. Define the post zero agency horizon as the first nonzero horizon where steering coupling falls to the null threshold or below. The detector focuses on τ greater than

zero to avoid instantaneous alignment artifacts.

8. Leverage gap and detector score. Define a leverage gap at horizon τ as the difference between prediction coupling and steering coupling at that horizon. The detector score should increase when prediction remains high while steering is below threshold. This score is designed to highlight readable but unreachable horizons.
9. Readability requirement. Choose a readability threshold η determined by domain needs and stakes. Define the candidate warning set as the horizons τ where prediction coupling is at least η. This prevents warnings in regimes where the future is not readable and the honest output should be uncertainty.
10. Persistence rule for stable triggering. To prevent single spike triggers, impose a persistence requirement. Choose integers m and n with $1 \leq m \leq n$, and trigger only if the warning condition holds for at least m horizons within a sliding window of length n, or for a contiguous block of at least m horizons. This reduces false alarms and improves trust.
11. Operational detection window. Define the operational detection window, ODW, as the set of horizons that survive the full filter. First, prediction must be readable, meaning prediction coupling meets the readability threshold. Second, steering must be collapsed, meaning steering

coupling is at or below the null threshold. Third, the condition must persist under the chosen persistence rule. The ODW is the practical region where the detector says the future is readable but direct steering is not operationally effective.

12. Lead time relative to failure acceleration. Define a failure event boundary for the domain, for example a future state falling into a failure set X_fail. Let Fail(τ) denote the failure rate by horizon τ across trials. Define a functional cliff as the first horizon where Fail(τ) exceeds a high level such as 0.99. A deployment requirement is that the ODW begins well before the functional cliff, providing usable lead time rather than a late receipt.
13. False positive accounting under null. Under null controls, measure the false positive rate as the fraction of runs where the ODW is nonempty or where triggers occur. Deployment requires that this rate remain low at the chosen β and persistence settings, otherwise the detector becomes noise and degrades response speed.

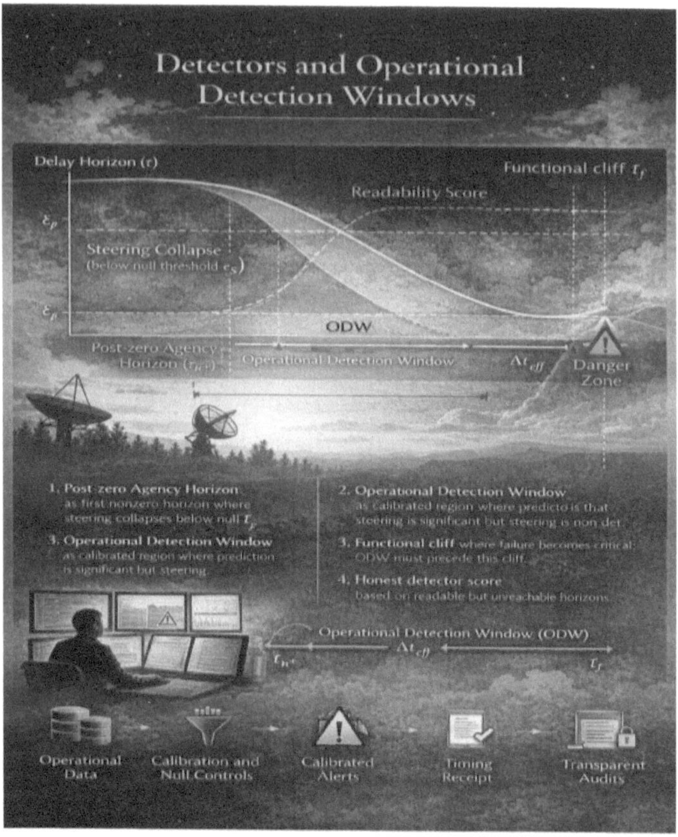

Detectors and Operational Detection Windows
Delay Horizon (τ)
Functional cliff τ_f
Readability Score
Steering Collapse
(below null threshold ε_S)
ODW
Post-zero Agency Horizon (τ_{H^+})
Operational Detection Window
Δt_{eff}
Danger Zone
1. Post zero Agency Horizon
as first nonzero horizon where steering collapses below null τ_f
3. Operational Detection Window
as calibrated region where prediction is significant but steering
2. Operational Detection Window
as calibrated region where predicto is that steering is significant but steering is non det.
3. Functional cliff where failure becomes critical: ODW must precede this cliff.
4. Honest detector score
based on readable but unreachable horizons
Operational Detection Window (ODW)
Operational Data
Calibration and Null Controls
Calibrated Alerts
Timing Receipt
Transparent Audits

Chapter 10

Ethics and Law Under VTM

The hearing begins the way weather begins now: quietly, automatically, with a polite light turning on. A wall-sized display wakes up and shows a simple timeline. No names yet. No faces. Just a horizontal line with dots and small labels.

00:00:00 — Event onset detected

00:00:02 — Risk elevated

00:00:04 — Notification delivered

00:00:07 — User acknowledgement registered

00:00:11 — Intervention permitted

00:00:13 — Intervention executed

00:00:14 — Outcome locked

The room is full, but the room feels secondary. Everything important is already on the screen: the sequence, the delays, the gaps between one dot and the next. The clock has become the star witness.

In the front row sits a man who looks like he hasn't slept since last year. He holds his hands together like they're fragile objects. If you asked him what he did wrong, he could tell you in a thousand different ways, and each one would sound sincere. People like him always volunteer guilt, because guilt is something you can hold. Guilt is heavy. Guilt is human. Guilt feels like control.

Above him the timeline waits, perfectly calm.

The case is famous because it is ordinary.

A commuter train came into a curve too fast. Not wildly too fast—just fast enough. The safety net noticed. The safety net always notices. That is what it is for. The passengers felt a small sway, a gentle complaint of metal on metal, and then the world did what the world does when mass and speed stop agreeing.

No explosion. No melodrama. Just physics reclaiming the schedule.

The public argument began before the ambulances arrived. It began in the same place arguments begin now: in the space between "you were warned" and "what could you have done?"

Because he was warned. Everyone agrees on that. It says so right there on the timeline: notification delivered. Then acknowledgement registered.

So why did it happen?

The first speaker doesn't do what speakers used to do. She doesn't pound a table. She doesn't perform outrage. Outrage is a cheap fuel now—fast burning, messy, unreliable. She speaks like a person reading the weather.

"We will determine responsibility by reconstructing accessibility," she says, and you can feel half the room stiffen, because that word is new enough to still sound like an excuse.

Accessibility.

Not "what was true." Not "what should have been done." Not even "what was possible" in the heroic, cinematic sense.

Accessibility: what information actually arrived, in time, in a usable form, for a real nervous system in a real body with real latency.

She gestures once, and the screen changes.

A second line appears beneath the first: a shaded band that begins at 00:00:04 and ends at 00:00:11.

"This," she says, "is the reaction window."

A third line appears: a thinner band inside the second, tighter, almost embarrassingly short.

"And this," she adds, "is the effective reaction window."

You can see it now. Between the arrival of the warning and the permission to intervene, there is a gap. Not a dramatic gap. Seven seconds. In old movies, seven seconds is enough time to whisper goodbye and turn the wheel and become a hero.

In real life, seven seconds is barely enough time for your mind to fully accept what your eyes are seeing.

The man in the front row looks up as if he is seeing the shape of his own sentence for the first time.

The speaker continues. "We will review the content of the notification."

The room is shown what he saw.

A gentle alert.

A small symbol.

A sentence that looks like it was designed by someone who fears emotion.

Stability anomaly detected. Please remain calm. Your cooperation improves outcomes.

Beneath it, a button.

ACKNOWLEDGE

That was all.

No "Brake now."

No "Pull the emergency lever."

No "You have four seconds before the curve."

Just calmness. Politeness. An instruction to cooperate.

The timeline shifts again. Another layer appears: physiological limits. A band labeled "typical recognition latency," another labeled "decision latency," another labeled "motor initiation."

The point is not to excuse him. The point is to stop pretending he was a machine.

The old idea of responsibility assumed something that never existed: instantaneous comprehension, instantaneous choice, instantaneous action. The old moral universe pictured a person standing inside time like a king standing on a balcony, surveying the road below, making decisions at leisure.

But in the Volumetric Time Model, you are not a king on a balcony. You are a moving point inside a finite volume, looking out through a narrow cone of access. Information does not arrive as truth; it arrives as signals filtered through distance, noise, attention, and the stubborn biology of your senses.

You can be surrounded by reality and still be blind to it.

You can be present and still be powerless.

The speaker does what all good futurists do: she makes the strange thing ordinary.

"In earlier centuries," she says, "we judged intent as if the mind were a clear room with a single door: you either knew or you didn't. You either chose or you didn't. Today, most harm occurs in conditions where the door is a hallway with a thousand mirrors."

She pauses, as if letting the metaphor settle into the floorboards.

"People are being held responsible for outcomes that were never accessible in time."

The audience murmurs. Some are relieved by the sentence; some are offended. A few feel the sudden, dizzy fear that comes when your moral reflexes are revealed to be habits.

Because the moral reflex is simple: if something bad happened and someone was nearby, that person must have failed.

It is a prehistoric instinct. It made sense when the world was small and feedback was immediate. If the fire went out and you were holding the stick, we could all point at you. If the animal got away and you were holding the spear, we could all blame you.

But the modern world is not a campfire. It is not a spear. It is layers of delayed signals, buffered permissions, ambiguous alerts, thresholds set by invisible committees of optimization. The harm is often not in what you did. The harm is in what you were allowed to reach.

The screen shows another fact: the permission to intervene did not arrive until 00:00:11.

"Why," the speaker asks, "was intervention not permitted at 00:00:04?"

A second person stands. He represents the manufacturer of the safety net. He looks polished in the way people look when their job is to appear harmless. He says what he has been trained to say.

"We prioritize stability," he begins. "False interventions can be disruptive. We require confirmation to reduce unnecessary action."

In other words: delay is a feature.

We built a world where action is gated by permission, and permission is delayed by design, and then we blame the individual for not acting quickly enough.

This is the kind of thing that would have made philosophers argue for a century. But in the VTM frame, it becomes brutally concrete. Responsibility is not a metaphysical debate. Responsibility is an audit of access.

What could the person actually know, in time?

What could they do, given the delays imposed on their options?

What fraction of the future was reachable through their cone of access?

And here we hit the uncomfortable truth that sits beneath so much modern life:

The most powerful force is not violence.

It is timing.

The speaker gestures again. The timeline zooms in to show a set of micro-delays most people never think about.

The warning arrived at 00:00:04, but it arrived while the man's eyes were already tracking another signal—the track ahead, the speed reading, the habitual scan of passengers. The alert was placed in the lower corner of his view, like a polite cough.

Acknowledgement was registered at 00:00:07. That looks like consent. That looks like participation. That looks like "he saw it."

But the panel does not treat acknowledgement as comprehension. The panel treats it as what it often is: reflex.

A finger pressing a button to make a light go away.

A muscle memory trained into the human animal by a thousand prior prompts.

You can acknowledge something without absorbing it, the way you can nod at a stranger without hearing their name.

The room watches the replay. The man presses ACKNOWLEDGE the way you swat a mosquito. Three seconds.

On paper, that looks like agency.

In lived experience, it is something else: compliance.

The speaker is careful with her language. She does not say the man is innocent. She does not say the manufacturer is guilty. She does not offer a villain to satisfy the ancient hunger for punishment.

Instead she offers a new question:

"Who controlled the clock?"

The room goes quiet because everyone understands the direction of the knife.

In the Volumetric Time Model, the future is not a blank page waiting for your pen. The future is part of the volume you are moving through. Your choices are vectors that select trajectories that already exist as admissible paths.

The only thing that changes, from moment to moment, is what becomes accessible in time to steer you.

And that is why the ethics of a volumetric universe become the ethics of access.

If you shrink someone's reaction window, you shrink their moral world. You shrink what they can be expected to do. You shrink what "choice" can honestly mean.

And if you do it politely—if you do it with a calm alert and a gratitude sentence and a button labeled ACKNOWLEDGE—then you can make the person feel responsible while quietly removing the ability to respond.

The panel calls the next item: a comparison case.

Another timeline appears. Another event, different domain.

A young woman with severe allergies eats a packaged meal at a stadium. She scanned the label. She checked. She did what she was told to do: be a responsible consumer in a world of infinite options.

The package did not lie. It simply did not say what mattered in time.

The allergen information existed, somewhere, in a database that could have been reached with the right query, the right connection, the right patience. It was true. It was recorded.

But it was not accessible to her, in the moment, in a form she could use.

The next dot on the timeline is cruel in its simplicity:

00:00:00 — Exposure

And then, almost immediately:

00:00:09 — Symptom onset

Ethics begins to look less like a sermon and more like physics.

The room is learning the central lesson of the ending of this book: the moral universe is shaped by latency.

Not your feelings about latency.

Not your opinions.

Latency as a measurable property of reality.

And once you see it, you start seeing it everywhere: in medicine, in finance, in relationships, in politics, in personal regret.

The speaker returns to the first case and says something that sounds mild but is explosive.

"We will not treat information that exists as equivalent to information that was accessible."

A few people shift uncomfortably. This sentence threatens entire industries built on the gap between what is technically available and what is realistically reachable.

It threatens the ancient tactic of power: to hide behind availability.

We gave you the notice. It was on the label. It was in the terms. It was in the fine print. It was in the third menu. It was in the help section. It was in the archive.

It existed.

But existence is not access.

The timeline waits.

The clock does not care about excuses.

The speaker looks at the room and gives them the thesis in plain language, the way a good physicist will, when the moment comes to stop being clever.

“Accountability migrates,” she says, “toward whoever controls the timing.”

And for the first time in the hearing, the man in the front row stops looking like a villain and starts looking like what he has been all along: a human being inside a delayed world, holding a button that looked like choice.

The screen does not blink. It does not apologize. It does not take sides. It just keeps time.

After the line *Accountability migrates toward whoever controls the timing*, the room exhales the way people exhale when they realize the villain might not be a person. Some look relieved. Some look cheated. A few look angry, because anger prefers a face.

The speaker lifts a hand and the timeline collapses into something else: a map.

Not a street map. An access map.

A human silhouette appears—generic, almost insulting in its neutrality—surrounded by a halo of colored zones. Close to the body, the colors are bright: signals the person could actually receive and use. Farther out, the colors fade into gray: things that existed, that were true somewhere, that would be discoverable if you had hours and perfect attention and ten extra hands.

But you don’t have hours. You have seconds.

And you don’t have ten extra hands. You have two, and one of them is usually holding something that buzzes.

“This,” the speaker says, “is what we mean by accessible.”

She doesn't define it with a sentence. She defines it by showing the audience what it looks like when reality is larger than reach.

The halo brightens around the commuter train case. A small red wedge appears at the lower edge of the silhouette's field of view: the location of the alert. Tiny. Polite. Positioned like a courtesy, not an emergency.

The gray zone expands where you expect it to expand: the manual, the training guide, the deeper diagnostic readout, the buried option that would have allowed immediate braking—if it were available without permission, and if the warning had carried instructions instead of gratitude.

On the far right of the display a simple phrase appears:

TRUE ≠ USABLE

It sits there like a cold coin.

The speaker turns to the manufacturer's representative again—not with rage, not with a prosecutor's theater, but with the neutral curiosity of a physicist who has just found a new variable.

"You said you delayed intervention to prevent false action."

"Yes," the representative answers quickly. "We had to minimize unintended disruptions."

"Minimize disruptions," she repeats, tasting the phrase.

Then she does what good futurists do: she time-travels without leaving the room.

"In the nineteenth century," she says, "when railroads expanded across continents, towns discovered that 'noon' wasn't a universal thing. Each city had its own local sun, its own local time. It was charming until it became deadly. Trains don't care about charming."

A few people smile, recognizing the story. It's a classic for a reason: humanity learned that when technology stretches space, you must reinvent time or you will crash into your own confusion.

"We invented time zones," she continues, "because we needed a shared clock."

She pauses.

"Now we need something else."

She gestures at the access map.

"We need shared standards for reach."

Because the modern world has done something quietly radical: it has separated truth from usability, and then pretended that was fair.

This is the great moral trick of the age. You can hide behind availability. You can say, with perfect legal calm, *It was there. You could have known.*

And then you can build a maze between a person and the knowledge they would have needed in time.

The speaker's voice stays calm.

"When you place critical information behind delay," she says, "you are not informing. You are decorating responsibility."

The phrase lands hard. Decoration is what you do when you want the appearance of a thing without the substance. A fake plant. A printed bookshelf. A button that does nothing but record obedience.

The man in the front row stares at the screen like it is showing him his own spine.

A second case appears, and the room shifts again. This one is not about a train. It's about money, which means it is about survival with fewer sirens.

A young couple tried to rent an apartment. They had jobs. They had savings. They had letters from previous landlords. They had, by older standards, what any reasonable person would call proof.

They were rejected.

The rejection arrived instantly, smoothly, with an almost cheerful efficiency.

The timeline shows:

00:00:00 — *Application submitted*

00:00:01 — *Score computed*

00:00:01 — *Decision delivered*

00:00:01 — *Appeal offered*

The appeal offered. That's the part that used to soothe people. The existence of an appeal makes the rejection feel civilized.

But the panel zooms in and shows the real timeline:

00:00:01 — *Appeal offered*

+14 days — *Appeal reviewed*

+30 days — *Outcome irreversible*

In other words: the appeal is a story you tell to make yourself feel humane while the clock carries out the real decision.

The couple does not need an explanation in two weeks. They need a home now.

The speaker turns to the audience.

"The old moral language would ask: 'Did they do something wrong?'"

She shakes her head gently.

"The correct question is: 'What could they realistically change inside the window where change mattered?'"

The access map appears again. The couple's halo is bright at the edges: they can read the rejection, they can click the appeal, they can write a heartfelt paragraph, they can upload documents.

And then the halo fades, because the factors that produced the score are upstream, hidden behind layers of proprietary logic, and even if the couple knew the precise reason, they could not repair it inside the window where the decision had teeth.

They are offered a lever that is connected to nothing.

In a volumetric universe, this is not merely unfair—it is structurally dishonest.

Because what makes an action meaningful is not the sincerity behind it. It is coupling. The world must answer your motion while it still matters.

If the answer arrives after the branch point—after the point where the possible paths narrow—then your action becomes performance.

You don't have agency. You have theater.

The speaker clicks again, and the audience is shown the rejection message the couple received. It is beautifully written. It thanks them for their interest. It values their time. It praises their patience.

And it is useless.

Someone in the back finally says what the room has been holding.

"So what are you saying? No one is responsible?"

The speaker smiles—not because the question is funny, but because it is familiar. It is the same reflex people had when they first learned that the Earth was not the center of the cosmos. If we are not central, then who is in charge? If the old story is false, do we have to live in chaos?

"No," she says softly. "I'm saying responsibility has been misassigned."

She points at the access map again.

"When you shrink someone's usable window, you shrink what you can honestly demand from them. If you want to assign blame, you must first assign clocks."

A new layer appears on the screen: the *clock owners*.

This is where the hearing becomes dangerous, because the audience begins to see the world in a way that cannot be unseen.

Clock owners are not necessarily tyrants. They can be well-meaning engineers, cautious policymakers, risk managers, product designers, legal teams. They can be ordinary people who never intended harm.

But they are the ones who decide:

- How early the warning arrives
- How clear the warning is
- Whether the person is given a direct action or a polite acknowledgement
- Whether the lever is real or symbolic
- Whether the appeal window exists while it still matters

These decisions are often framed as "safety," "stability," "reducing false action," "improving experience."

But in the VTM frame, they are moral architecture.

The speaker shows a final case, smaller, more intimate, and because it is intimate it hurts more.

A man receives a message from his sister: *Dad is getting worse. Please come.*

He reads it. He does not answer immediately. He is in a meeting—no, not that word—he is in the middle of a day that has already been scheduled around him. He tells himself he will reply in an hour.

The timeline appears. It is almost too petty to dignify with graphics.

00:00:00 — *Message received*

+1 hour — *Reply drafted*

+3 hours — *Call attempted*

+3 hours — *No answer*

+5 hours — *Update received: too late*

Some people in the room flinch because they recognize it. This is not technology anymore. This is life. This is regret.

The speaker doesn't moralize. She doesn't say "be better." She doesn't sell self-help. She simply points to the cruel reality that the VTM makes visible:

Your life is full of branch points you do not recognize until after you pass them.

Sometimes the branch point is a curve in a train track. Sometimes it's a contract clause. Sometimes it's a human heart that had a window and then closed.

And the tragedy is not that you were evil. The tragedy is that you were late.

This is why the ending of this book cannot be metaphysical comfort. It has to be operational truth.

A universe can be fixed in its admissible paths and still feel open inside the windows where your actions actually couple to outcomes. But if the windows are shrunk—by design, by delay, by noise, by politeness—then life becomes a set of acknowledgements.

You clicked. You agreed. You participated.

Nothing changed.

The speaker returns to the train case one last time, but she no longer looks at the man in the front row. She looks at the designers who built the warning.

"What," she asks, "was the purpose of that button?"

The representative hesitates.

"It confirms receipt," he says.

"Receipt," she repeats.

Then she says a sentence that is so simple it feels like it should have been obvious from the beginning, and that is why it is frightening:

"Receipt is not consent."

She taps the screen again. A new standard appears—proposed, not yet law, but already inevitable because it has the bluntness of physics.

Informed action requires:

1. Timely delivery
2. Comprehensible meaning
3. A real lever inside the effective window

Anything else is paperwork for the conscience. A ritual of liability transfer disguised as user choice.

The room is silent now, the way a room is silent when it realizes it has been living inside a trick.

The speaker concludes this segment with the kind of forward-looking pivot, the moment where history turns into tomorrow.

"We are entering an era," she says, "where courts will not argue about metaphysics. They will audit access."

She gestures at the access map like it is a new kind of fingerprint.

"Because the central ethical question is no longer 'What did you intend?'"

She pauses, and the pause is the whole knife.

"The central question is: 'What could you reach in time?'"

And as those words settle, you can feel the audience beginning to understand the real ending of this book, the ending that will follow them out into the street:

If someone controls your timing, they control your moral world.

They can make you guilty without giving you leverage.

They can make you click and call it choice.

They can make you late and call it character.

The clock does not care. But we should.

Outside, the air feels too simple.

People step into the street as if leaving a planetarium. Their eyes keep trying to refocus on ordinary distances—traffic lights, faces, the dull choreography of pedestrians—but the thing they just saw follows them like an afterimage: a line with dots on it, a shaded band called *effective window*, a blunt phrase like a verdict:

TRUE ≠ USABLE.

It's hard to explain what changes in a person when you show them timing as a moral object.

Nothing dramatic happens. No one falls to their knees. No one confesses.

They just start noticing the delays.

They notice how often they are asked to *acknowledge* things that offer no lever. They notice how frequently "choice" is presented at the exact moment when choice can no longer steer anything. They notice that the world has become very good at placing responsibility downstream from the last meaningful branch point.

And once you see that, you begin to understand a strange truth: the modern world is not built out of laws. It's built out of windows.

In earlier centuries, the critical invention was the shared clock. Time zones. Train schedules. The synchronization of distant places into a single choreography. You could not run a continent on local suns. Steel does not negotiate with romance.

Now the next invention is arriving, quietly, without parades: access zones.

Not time zones—access zones.

The idea is simple enough to be dangerous: we will stop treating information that exists somewhere as equivalent to information a person could use in time. We will stop pretending that late truth is a form of empowerment. We will stop rewarding designers who hide behind availability.

This shift doesn't require grand philosophy. It requires a new kind of accounting.

A latency budget.

Every warning has a budget. Every choice has a budget. Every consent prompt has a budget. Every "appeal" has a budget. The budget is measured in seconds, minutes, days—whatever scale matters for the consequences at stake.

If the budget is spent, the act becomes decoration.

And here is where the Volumetric Time Model changes ethics from a sermon into geometry.

Imagine two identical universes. Same people. Same trains. Same meals. Same contracts. Same storms. Same hearts.

The only difference is timing.

In Universe A, a warning arrives early enough to be acted on, and action is permitted without delay. A lever is real. When you pull it, the world answers while the branch is still open.

In Universe B, the warning arrives late, or action is gated until the branch closes. The lever is theater. When you pull it, you feel movement, but the world does not move with you.

In the old moral language, we would judge the outcomes as if they were evidence of character. In the VTM language, we judge them as evidence of coupling.

The difference between a responsible act and a tragic one may be nothing more than four seconds of withheld permission.

That is not a poetic statement. It's an engineering statement. And engineering, as always, becomes ethics the moment it touches human consequence.

This is why the phrase "clock owners" makes people uneasy. It sounds like conspiracy. It sounds like a cartoon villain with a pocket watch.

In reality, clock owners are everywhere and almost never malicious.

They are the ones who decide that a warning should be "gentle," because panic is bad for engagement metrics.

They are the ones who decide that action must be confirmed, because false action is "disruptive."

They are the ones who decide that an appeal must exist, because appeals make decisions feel fair, even if the appeal cannot alter the outcome in time.

They are the ones who decide how many steps it takes to turn something off.

How many menus you must traverse to escape.

How many seconds the interface will pause before it allows you to opt out.

How long the process will "process."

To see the cruelty hiding in politeness, you do not need a catastrophe. You can watch something smaller, like a smoke alarm.

In one house, the alarm screams and the door unlocks. The lights turn on. The route is illuminated. You are given a clear instruction: *Move now.*

In another house, the alarm whispers: *Air quality anomaly detected. Please remain calm.* A button lights up: ACKNOWLEDGE. The door remains locked until the device confirms the reading. The device confirms the reading only after a second sensor agrees. The second sensor is delayed because it is "calibrating."

The house burns politely.

No one intended harm. Everyone designed for stability.

But the consequence is written in the clock.

When people ask why this matters, they usually mean: why should timing belong in ethics at all?

The answer is: it already does. We simply hid it.

Timing is the invisible substance of fairness. A promise delivered after the deadline is a joke. A warning delivered after impact is a receipt. A choice offered after the branch point is a performance.

When you remove access, you don't always remove truth. Often you increase truth. You create a world where everything becomes legible—scores, forecasts, risk labels, probability estimates. The future begins to look clean.

And that is the trap.

Because there is a peculiar curve that emerges again and again in both machines and human lives: as meaningful influence collapses, predictability can rise.

The closer you are to helplessness, the more "obvious" the future can become.

You see this in markets. You see this in medicine. You see this in relationships. You see this in disasters that feel inevitable only in retrospect.

There is a sick kind of clarity that arrives when the branch points close.

In a volumetric universe, this clarity is not magic. It is a signal that the cone of accessible paths has narrowed. Fewer trajectories remain reachable. The future becomes easy to guess because there are fewer options left.

This is why late certainty should scare you.

When the world starts making perfect sense, check your levers.

That sentence should be printed on every "smart" screen.

It explains why so many people feel as if they are living inside fate. They are not imagining it. They are feeling the geometry. They are sensing the narrowing of accessible trajectories. They may still receive beautiful predictions—clean forecasts, confident recommendations, risk flags that arrive like gospel—but the ability to steer has already been throttled below the threshold where it matters.

You are told what will happen.

You are not given the power to change it.

This is forecasting without power, translated into the language of daily life.

And once you see it, you begin to understand a new kind of cruelty that has become common: moralization of the powerless.

We scold people for not acting differently in conditions where action had been drained of coupling. We demand "better choices" after the world has been built to ensure choices arrive as acknowledgements.

A person misses a payment and gets buried in fees. The warning existed. It was sent. It was recorded. The person was "notified."

But the warning was delivered at 2:17 a.m. beneath twenty other prompts. The only immediate lever offered was "OK." The option to change the payment schedule required a phone call during a narrow window of hours. The phone call required a hold time longer than the person's lunch break. The hold time required patience that costs money. The person runs out of time, then is blamed for irresponsibility.

You can repeat this story in a thousand domains. The details change. The clock does not.

So the emerging standards of access—these access zones—do not begin with lofty declarations. They begin with three blunt questions:

1. Was the signal delivered in time?
2. Was it comprehensible to a real person?
3. Was a real lever available inside the effective window?

If the answer to any of these is no, then calling the outcome "choice" is dishonest.

Notice what this does to consent.

We used to treat consent as a checkbox. A signature. An acknowledgement. The ritual was enough to transfer liability. But if consent is to be real in a volumetric universe, consent must be coupled to accessible action.

A human being cannot consent to what they cannot reach in time.

Consent requires a lever that exists before the branch point closes.

Consent requires reversibility while reversal still matters.

Otherwise consent is paperwork for the conscience.

And this is the point where some people become nervous—not because they disagree, but because they see the consequences.

If we take access seriously, whole industries built on delayed levers begin to look like confidence tricks. If we audit timing, "compliance" becomes less impressive. If we measure effective windows, a lot of polite interfaces start to resemble traps.

That is why access zones will not arrive easily. There will be resistance, and not only from obvious places. Even well-meaning designers will resist, because it's uncomfortable to admit that stability can be cruelty and politeness can be harm.

And yet the physics does not care what is comfortable.

A lever that arrives late is not a lever.

A warning that cannot be acted on is not a warning.

A choice that cannot steer is not a choice.

The deeper implication is even more unsettling: if responsibility migrates toward whoever controls timing, then the moral center of gravity moves upstream. It moves toward the people who allocate delay budgets, who decide which options are immediate and which are buried, who decide whether a person's "no" is honored as an actual branch or recorded as a preference and ignored.

That means responsibility becomes less theatrical and more architectural.

It becomes less about punishing individuals and more about auditing design.

And that, in turn, introduces a new fear:

What if the audit itself becomes a weapon?

If you can prove that someone had no access, you can absolve them—but you can also sideline them. You can categorize them as incapable of consent. You can decide, gently, that their levers should be removed for their own stability. You can "protect" them by shrinking their world until nothing dangerous is reachable.

Access audits can free people from unfair blame.

They can also become a polite mechanism for taking options away.

The street outside is full of ordinary life—delivery vans, teenagers laughing, someone arguing into a headset, a dog pulling a tired owner toward a tree—but underneath it, a new question has begun to glow in the minds of those who saw the timelines:

If timing is power...

...who will be allowed to own it?

The fear arrives in the nicest possible outfit.

Not with sirens. Not with black vans. Not with a boot on a neck.

It arrives as a gentle reform proposal, written in clean language with reassuring verbs: *protect, support, streamline, prevent harm.*

Because once you build a world where responsibility depends on what was reachable in time, you also build a world where the next move is obvious:

If someone repeatedly fails inside a narrow reaction window, why not "help" them by removing the window entirely?

Why not take away the lever—so they can't be blamed for not pulling it?

And that is the moment when a tool meant to stop unfair blame begins to mutate into something else: a polite method of shrinking lives.

The panel knows this. The speaker knows this. That's why she doesn't end the hearing with celebration. She ends it with a warning that feels almost too calm to be a warning.

"Access audits," she says, "can liberate people from false guilt."

She lets the sentence sit, like a good scientist letting the first result land.

"And they can also be used to reclassify people out of power."

Reclassify. That word hits the room differently. It doesn't sound like violence. It sounds like paperwork. Like a clerical change in a database. Like a harmless rearrangement of categories.

But categories are fate when the world runs on gates.

The screen changes again. This time the timeline is not an accident. It's a life.

A young man applies for a license to operate heavy machinery. He has the hours. He has the training. He has the desire. The job pays well. The job would change his family's trajectory.

He is denied.

Not because he failed an exam. Not because he broke a rule.

Because his access profile is flagged.

The panel shows the logic in the blunt language of modern decision-making:

- Reaction latency: elevated
- Attention variance: elevated
- Risk tolerance mismatch: elevated
- Prior incident probability: elevated
- Intervention compliance: inconsistent

None of these are lies. None of them are mystical. They are measurements derived from data. And yet the young man's face on the screen—his stunned, offended, frightened face—makes the room feel suddenly ancient, because the moral question is ancient:

What do you do with a person once you can describe them as a risk curve?

The speaker leans forward slightly.

"This," she says, "is the temptation."

The temptation is to treat the audit as destiny.

To treat the profile as identity.

To treat a measured limitation of access as a reason to remove options.

And to do it with a smile.

The audience is shown the denial message the young man received. It is beautifully phrased. It thanks him. It praises his initiative. It suggests alternative opportunities "better aligned with his safety profile." It offers resources. It offers coaching.

It offers everything except the lever he asked for.

What makes the message cruel is not its tone. It's the timing. The decision is immediate and irreversible, while the appeal—again, always the appeal—moves slowly through a window that cannot affect the branch point.

The panel shows a shaded band: effective window.

It is already closed.

This is the new form of exclusion: not punishment, but prevention. Not condemnation, but "care." Not force, but "alignment."

And because it is framed as safety, it becomes difficult to argue against without sounding as if you want people to get hurt.

This is how the ethical trap tightens. The tool was built to stop unfair blame, but the moment you can quantify reach, you can also quantify who should be allowed to steer.

The speaker steps away from the screen and walks—slowly, deliberately—to the side of the room where a smaller display waits.

She taps it, and the audience is shown something that looks like a passport.

A simple card.

Not with a name, not with a nationality, but with a few numbers and colored bars.

ACCESS BAND: A
REACTION WINDOW: 2–4 seconds
REVERSAL PRIVILEGE: enabled
HIGH-IMPACT ACTIONS: permitted

Then another card:

ACCESS BAND: C
REACTION WINDOW: 6–12 seconds
REVERSAL PRIVILEGE: limited
HIGH-IMPACT ACTIONS: conditional

Then another:

ACCESS BAND: E
REACTION WINDOW: variable
REVERSAL PRIVILEGE: disabled
HIGH-IMPACT ACTIONS: restricted

The room shifts the way a room shifts when it realizes the future is not science fiction—it's a policy draft.

"Do you see it?" the speaker asks.

She is not asking dramatically. She is asking like a physicist pointing at a graph. *Do you see the curve?*

Because once this kind of access passport exists, it will creep.

It will start with obvious high-risk domains: aviation, medicine, heavy transport. Few people will object. It will sound reasonable.

Then it will move into finance: "to protect you."

Then into education: "to support you."

Then into relationships: "to reduce harm."

Then into speech: "to improve outcomes."

And at each step, it will be justified with the same logic: you are not being punished. You are being protected.

But the end result is a quiet caste structure built out of timing.

The speaker returns to history again, because history is where futurists go to keep themselves honest.

"In the past," she says, "power was often visible. It wore uniforms. It carried weapons. It declared itself."

She pauses.

"Today power is frequently expressed as delay."

Delay is harder to protest. You can't photograph it. You can't point at it the way you point at a bruise. It hides in customer support hours, in slow appeals, in hidden menus, in gates that ask for one more confirmation, one more proof, one more day.

Delay turns a right into a suggestion.

Delay turns freedom into a ritual.

Delay turns consent into a receipt.

The panel is not debating whether measurement is real. Measurement is real. The panel is debating what happens when measurement becomes the gatekeeper of personhood.

A woman in the audience stands. She has the voice of someone who has lived inside these delays.

"My mother has dementia," she says. "She forgets. She clicks things. She signs things. She gets scammed. Are you saying we should give her full access just because access is a right? Should we let her destroy herself?"

The question is sincere. It is the kind of question that makes abstract ethics collapse into tears.

The speaker doesn't answer with ideology. She answers with geometry.

“No,” she says. “I’m saying we should stop pretending the only options are full access or no access.”

She gestures, and the access passport changes. A new field appears:

REVERSIBILITY WINDOW

She explains it in a single sentence:

“Protection should be built out of reversibility, not removal.”

This is one of the key moves in the ethics of a volumetric universe: if you cannot give someone perfect reach, you can at least give them room to undo.

Because a reversible action keeps the future open.

An irreversible action closes branches.

And closing branches is the real harm.

This is why “care” becomes dangerous when it is implemented as restriction instead of reversibility. When you restrict, you remove possible trajectories. When you create reversibility, you preserve them.

The panel shows examples.

A bank transfer above a threshold triggers a delay—not a denial. The money does not vanish. It waits in a reversible state long enough for a second signal to arrive: a phone call, a trusted contact, a simple confirmation in plain language.

A legal signature is not accepted instantly; it enters a cooling period with a visible countdown and a clear “undo” lever.

A high-impact medical consent is not buried in a paragraph; it is presented as a short sequence of plain statements, each with an “I understand” that actually means something, because it can still change the outcome in time.

This is not about making life slower. It's about putting delay where it restores agency rather than stealing it.

And here the speaker does something that feels almost like a magic trick, except it's only logic:

She flips the idea of delay.

Delay can be oppression.

But delay can also be mercy.

A delay that arrives before the branch point closes is protection.

A delay that arrives after the branch point closes is theater.

So the ethics are not "delay is bad." The ethics are: **who gets to place delay, and where does it sit relative to the branch points?**

The room begins to see a new kind of civil right emerging, not written in lofty language, but in engineering constraints:

The right to an effective window.

The right to a lever that arrives while it still matters.

The right to reversal when the stakes are high.

The right to clarity that is timed to action.

It sounds almost trivial until you notice how much of your life is shaped by the absence of these rights.

The speaker returns to the earlier phrase on the screen—TRUE ≠ USABLE—and modifies it.

A new phrase appears beneath it:

USABLE MUST BE GUARANTEED

And then, like any good scientist who knows the danger of slogans, she adds the fine print, not as legal escape, but as a physical constraint:

"Guaranteed within domain-specific reaction limits."

In other words: if the decision has consequences that unfold in seconds, the access must arrive in seconds. If the decision unfolds over weeks, the access can arrive slower, but it must arrive early enough to influence outcomes.

The room is very quiet now, not because it is convinced, but because it is calculating the scale of what this would change.

It would change how products are built.

It would change how contracts are written.

It would change what consent means.

It would change how blame is assigned.

It would change how "help" is implemented.

And it would force a new question that the panel has been avoiding until now, because it is the hardest question in the room:

If we make access auditable, who gets to run the audit?

Because the audit itself—this beautiful tool for fairness—can become a gate.

A gate that decides whether you were "capable" of a right.

A gate that decides whether you are allowed to steer.

A gate that can be used to justify restriction with a smile.

The speaker looks at the audience with the calm intensity of someone who knows she is pointing at the next moral battlefield.

"We need," she says, "standards for access."

She pauses.

"And we need standards for the standards."

That second line is where the future hides.

Because without it, the audit becomes just another lever—owned by someone else, timed to someone else's convenience, used to move people quietly into categories that sound like care and feel like cages. The phrase **standards for the standards** hangs in the air like a second moon.

Everyone in the room understands the problem instantly, because it's the same problem humanity has had every time it invented a powerful measurement:

The ruler can be bent.

The scale can be tuned.

The thermometer can be placed in the shade.

And the moment a measurement becomes the gatekeeper of rights, careers, money, safety, and dignity, people stop arguing about truth and start fighting over calibration.

The speaker doesn't dramatize it. She doesn't need to. The audience is already running the movie in their heads: a future where your access passport determines what you can do, and the people who issue passports insist their process is "objective," and the people who are denied are told to appeal into a window that closes before it can matter.

The speaker taps the display again. The access passports fade, replaced by something that looks almost too mundane to carry this much moral weight:

A seal.

Not a national seal. Not a corporate logo. A certification mark—clean, geometric, neutral as a physics diagram. Under it, a short line:

ACCESS COMPLIANT — EFFECTIVE WINDOW VERIFIED

The room collectively leans forward. They recognize the shape of this idea. It's not new. Humanity has done this before.

"In the early days of electricity," the speaker says, "people died in their homes because wires were installed by whoever happened to have confidence and pliers."

A few smiles. A few grim nods.

"We solved that," she continues, "not by arguing about metaphysics, but by building standards—clear constraints. Wire gauge. Insulation. Breakers. Grounding."

She pauses.

"And then we built standards for those standards."

She doesn't say the names of organizations. She doesn't need to. The point is the pattern: you don't tame a dangerous technology by wishing for virtue. You tame it by making the failure modes legible and the constraints enforceable.

Now she turns back to access.

"Access is a safety-critical property," she says.

Some people flinch at the phrase because it sounds like a category expansion. Safety-critical used to mean bridges and airplanes. Now it's being applied to menus, alerts, consent prompts, appeals, and the invisible timing decisions that shape lives.

But the speaker is right, and the room knows it, because the case studies are already stacked in their memory like a deck of cards: the train curve, the allergic reaction, the housing denial, the life message that arrived in time to be read but not in time to be answered.

"What kills people," she says, "is not only impact. It is often the removal of usable windows."

She taps again. The screen shows a new graphic: a simple rectangle labeled **Domain**. Inside are categories:

- Transport
- Health
- Finance
- Housing
- Employment
- Public safety
- Personal data
- Communication

Under each category, a number appears:

Minimum Effective Window

Transport: *seconds*

Health: *minutes to hours*

Finance: *hours to days*

Housing: *days to weeks*

Employment: *days to weeks*

Public safety: *seconds to minutes*

Personal data: *days*

Communication: *minutes to days*

Nothing here is mystical. It's just a sober reminder that different consequences unfold on different clocks, and the clock you need is determined by how fast your life can change.

Then she adds the next line, and this is where the ethical architecture becomes clear:

If you offer a lever, you must guarantee an effective window large enough to use it.

The manufacturer's representative clears his throat. He knows what's coming, and he doesn't like it.

"That would require disclosure of internal designs," he says. "It could compromise proprietary—"

The speaker raises a hand gently.

"No," she says. "It requires disclosure of timing."

Timing, not trade secrets. Not code. Not the hidden details of how their devices make decisions. Just the measurable, auditable fact of when signals arrive, when actions become possible, and whether a person had a meaningful chance to steer.

She looks at him with a kind of scientific patience.

"If you want to keep your internal mechanisms private," she says, "that is your right. But you cannot keep the clock private."

There is a quiet shock in the room, because people understand that in a world run by gates, the clock *is* the mechanism.

The speaker continues.

"We already do this with food," she says. "We don't ask every manufacturer to reveal their recipes. We ask them to label allergens clearly. We ask them to declare hazards in plain language. We ask them to do it in time—before the bite."

She taps the display and a new phrase appears beneath the certification mark:

DISCLOSE HAZARD, DISCLOSE WINDOW, DISCLOSE LEVER

Then she adds a fourth, smaller line:

DISCLOSE REVERSIBILITY

Because this is where the trap can be avoided.

If access profiling becomes a caste passport, the antidote is not pretending measurement doesn't exist. The antidote is insisting that protection is built from reversibility rather than removal—and that reversibility is guaranteed when stakes are high.

"High-impact actions," she says, "should not be instant."

A few people look surprised. They expected her to demand speed everywhere.

"Instant decisions feel modern," she continues, "but they often produce irreversible harm before a human being can even recognize the branch point."

She points to the reaction window bands and the effective window bands again.

"There is a difference between preventing harm and preventing choice," she says. "Reversibility protects without erasing trajectories."

Now the hearing turns into a design session, which is where.

The point where you stop describing the world and start proposing the next version of it.

The speaker lays out what she calls an **Access Charter.** Not a constitution, not a sermon—an engineering brief for a civilization that finally admits timing is power.

Charter Principle 1: The Right to an Effective Window

If a decision has real consequences, the person impacted must receive timely, comprehensible information early enough to alter the outcome.

Charter Principle 2: The Right to a Real Lever

If you present a choice, it must couple to the outcome within the window. Otherwise, it must be labeled as informational only.

Charter Principle 3: The Right to Reversal

High-impact actions must include a cooling period or undo mechanism when feasible, scaled to domain risk.

Charter Principle 4: The Right to Plain Meaning

Critical prompts must use plain language. Not polite fog. Not legal poetry. Meaning is part of access.

Charter Principle 5: The Right to a Timing Receipt

Every consequential decision must produce a record of when signals arrived, when actions became available, and when branch points closed.

Timing receipt.

That phrase lands hard because it changes the power dynamic. It turns invisible delays into objects you can point at. It makes the clock speak in a way that can be carried into the street.

A woman in the audience raises her hand. "If we record timing receipts for everything," she says, "won't that be surveillance?"

A perfect question. A necessary question.

The speaker nods. "Yes," she says. "If we build it badly."

Then she does another move: she takes a problem that sounds purely political and makes it feel like physics.

"We need to distinguish between *records that trap* and *records that liberate*," she says. "A timing receipt should not be a diary of your life. It should be a diagram of the interaction."

A diagram, not a confession.

She proposes a constraint: the receipt is minimal and domain-specific. It records the clock and the lever—not the content of your thoughts, not your private messages, not your identity in rich detail. It's like a black box recorder in a plane: it doesn't exist to shame a pilot. It exists to reconstruct whether the machine and the environment allowed a human to respond.

That analogy flips the emotional temperature. People understand black boxes. They're not there to moralize. They're there to reveal failure modes and prevent repeats.

The manufacturer's representative tries again.

"But who controls the certification mark?" he asks. "Who certifies the certifiers?"

Good. Now we are finally at the center.

The speaker's answer is simple and unsettling: no single entity should.

She proposes a structure that looks almost biological.

Multiple independent certifiers, with transparent methods, competing and cross-checking each other. Rotating auditors. Public test suites. Randomized spot checks. Openly published failure cases. A living catalogue of known timing tricks—delays disguised as stability, levers that move the interface but not the outcome, appeals that exist after the branch point.

In other words: the audit must be auditable.

And here the Volumetric Time Model gives the whole concept a deeper spine:

If the future is a finite volume and your lived life is a trajectory through it, then any process that consistently shrinks effective windows is not just "inefficient." It is trajectory theft. It narrows your reachable set.

That isn't rhetoric. It's geometry.

The speaker stands very still and says:

"We are not banning prediction."

She looks around the room.

"We are banning prediction masquerading as empowerment."

Because that is the true late-stage trick: you show people beautiful forecasts and call it agency. You let them watch the cliff approach in high resolution. You let them feel informed while quietly removing the levers that could change anything.

This is the shape of despair in the modern age: high legibility, low coupling.

The hearing is nearing its most dangerous moment, because something else becomes obvious once you accept these principles:

Some people will try to game the timing receipts.

They will deliver warnings early but make them incomprehensible.

They will provide levers that exist but are buried.

They will offer reversibility but attach penalties.

They will flood people with signals until signal becomes noise.

They will meet the letter of the standard while violating the spirit through overload.

The speaker anticipates it. She says, almost casually:

"Information overload is a form of delay."

That sentence should be carved into steel.

Because overload doesn't deny access by hiding truth. It denies access by drowning it.

You cannot use what you cannot parse in time.

So the charter adds one more principle, and it feels like the hinge the entire ending may swing on:

The Right to Signal Integrity

Critical signals must be distinguishable from noise. If you bury urgency under a mountain of trivial prompts, you have stolen the window.

The room is quiet again, but it's a different quiet. This is not the quiet of shock. This is the quiet of recognition—people seeing their own lives in the architecture.

The speaker turns back to the screen and brings up the train case one last time. She overlays the charter principles on the timeline.

The warning arrives: not compliant.

The lever: not real.

The permission: late.

The receipt: absent.

Then she says, softly, almost kindly:

"This was not a failure of character."

She pauses.

"It was a failure of window design."

And now the hearing has become something else: not a trial of a man, but a preview of a future where we stop blaming people for being human inside delayed reality.

But the last problem remains.

Who gets to hold the certification mark?

Who gets to decide what counts as access?

Who gets to own the standards for the standards?

The speaker looks at the room and gives them the only answer that survives long-term:

"No one," she says. "Not alone."

Then she adds the real warning:

"Because the moment a single hand owns the clock, it will become law."

And the timeline waits, as if it already knows.

The room does not erupt. It does not clap. It does not boo. It does what rooms do when a new kind of truth is placed on the table, it turns quiet, then it begins to negotiate.

People begin to talk in the careful tone that means they are already imagining the second order effects. They are not arguing about the train anymore. They are arguing about what happens when timing becomes a measurable obligation. They are arguing about what happens when access becomes a right that can be audited.

The certification mark still glows on the screen, a clean shape with a clean promise. Access compliant, effective window verified.

The speaker stands in front of it and watches the faces shift. She has seen this before, in other centuries, with other inventions. First comes the revelation, then comes the scramble to own the revelation.

A man in the second row raises his hand. He represents a consortium, a coalition, a group that exists because groups are how complex societies pretend to speak with one voice.

He says that the mark is a good idea. He says that standards are necessary. He says that safety demands uniformity. He says the word uniformity as if it tastes like relief.

The speaker nods. Then she says, gently, that uniformity is not the same thing as fairness.

Uniformity can become a mask. It can become a single dial that everyone pretends is objective, while the real world remains messy and unequal. It can become a way to simplify a complicated moral landscape until it fits inside a single checkbox.

The consortium man insists that multiple certifiers will confuse the public. He insists that people need one trusted mark, one authority, one place to look.

The speaker listens, then points at the timeline again. She points at the shaded band that represents the effective window. She points at the place where the band is narrow enough that it becomes almost invisible.

She says that confusion is a tax, but monopoly is a cage.

She says that if one hand owns the mark, that hand will eventually own the clock. Not because the hand is evil, but because incentives pull in that direction the way gravity pulls in a direction. The more power you centralize, the more the future begins to bend toward the needs of the center.

A woman near the back speaks next. She does not represent a consortium. She represents the kind of quiet expertise that comes from watching people lose arguments against time.

She says that the certification must be testable, not declarative. She says that it must be based on public trials, on transparent measurements, on repeatable procedures. She says that a mark without a test suite is a brand, not a guarantee.

The speaker nods again. Then she changes the screen.

The certification mark shrinks to one corner, and the rest of the display becomes a laboratory.

Not a room full of microscopes, not a fantasy of white coats and dramatic discoveries. A simple testing protocol, domain by domain, with timed prompts and timed levers, with human reaction limits treated as physical constants rather than moral failures.

The speaker calls it a public test suite. She says the words slowly, the way you say a phrase you want people to remember.

Public test suite.

She explains what it means in plain language. If an organization claims that its warnings are timely, the claim must be provable by replay. If it claims that its levers are real, the claim must be provable by outcome coupling within the effective window. If it claims that it offers reversibility, the claim must be provable by an undo that actually changes the state before the branch point closes.

The consortium man says that this will slow innovation. He says that companies will hesitate. He says that the world cannot afford delays in progress.

The speaker does not argue about progress. She argues about what progress means if it arrives as prediction without leverage. She says that innovation that produces beautiful forecasts while shrinking real windows is not progress, it is legibility layered on top of helplessness.

She says that in a volumetric universe, progress is measured by reachable futures, not by dazzling descriptions of futures.

The room goes quiet again because the phrase reachable futures does something people do not expect. It makes the moral argument feel like geometry. It takes the question of fairness and turns it into a question of space inside the volume, a question of how many trajectories remain open when a person acts.

Then the speaker shows the audience the first adversarial trick.

She does it the way a good scientist does it, by demonstrating rather than accusing.

On the screen a device is tested. A warning appears early. The warning is technically inside the minimum effective window for its domain. The device passes the timing requirement.

Then the speaker replays the warning again, and again, and again.

The warning is written in language that requires a trained mind to parse. The warning is wrapped in polite fog. The warning is filled with terms that sound safe but mean nothing actionable. The warning is technically early, but it is effectively late because meaning arrives too slowly.

The speaker says that this is one of the oldest timing tricks, to deliver early but withhold clarity.

She says that clarity is part of timing.

The room murmurs. Some people nod, some people look defensive, because everyone has seen this trick before. Everyone has lived it. You are told in advance, but you are told in a way that makes you depend on someone else to interpret the signal.

So the speaker adds another requirement to the test suite. Comprehensibility must be measured. Not by taste, not by literature, but by timed recognition trials with real people, with plain benchmarks, with an honest accounting of how long it takes a typical nervous system to understand what the signal implies.

The consortium man says that people vary. He says that comprehension is subjective.

The speaker agrees, then she says that gravity is also subjective if you are careless. She says that variation is real, but that is why the requirement is built as a band, not as a single point. A device does not need to be perfect for every mind. It needs to meet a standard for human limits that are well understood, and it needs to offer a path for those outside the typical band, such as longer windows, stronger reversibility, clearer levers.

Then she shows the second adversarial trick.

A lever is offered. It is present. It is obvious. It sits right there on the screen.

The lever does something. It changes the interface. It produces a satisfying visual confirmation. It makes the person feel the warm certainty of action.

But the outcome does not move.

The lever is a theater lever.

The speaker calls it ornamental control. She does not insult it. She does not laugh. She treats it like a failure mode, the way engineers treat a crack in a beam.

She says that ornamental control is worse than no control, because it trains people to confuse compliance with steering. It trains people to spend their attention and their trust on gestures that do not couple.

So the test suite adds another requirement. Outcome coupling must be demonstrated. If a lever is presented as an action, the lever must change the state in a way that can be verified in the timing receipt. If it cannot, the interface must label it as informational, not actionable.

The room feels colder now, not because the room is hostile, but because it is becoming honest.

A third person speaks, this one from the consumer side, from the side of the people who have always been told to read the fine print and be responsible. He says that companies will simply drown people in signals. He says they will meet timing and lever requirements by flooding the channel.

The speaker nods, and she says that overload is a form of delay, because signal integrity collapses when everything demands attention.

So the test suite adds another measure. Signal integrity must be audited. Critical signals must be distinguishable from noise. If a device produces too many prompts, too many banners, too many nudges, the device is not compliant, because it has turned the human mind into a clogged pipe.

The audience begins to see that the certification mark is not a decoration, it is a constraint.

But constraints provoke the next move, which is not technical at all. The next move is capture.

Capture is what happens when a standard becomes valuable. People do not need to be malicious to capture a standard. They simply need to prefer outcomes that benefit them, and to have the power to shape definitions.

The consortium man returns to the microphone and says that public test suites are fine, but that the thresholds must be practical. He says that the windows must be realistic. He says that too strict a standard will create instability, and instability harms everyone.

The speaker listens. Then she says that instability is a word people use to defend delay, because delay feels calm.

She says that in many domains, stability is not the same as safety. Stability can mean that harm happens quietly, predictably, and too late to prevent.

She says that the purpose of an access standard is not to make the world feel smooth. The purpose is to keep branch points reachable.

Then she does something subtle. She changes the screen to show a distribution. Not a single window, but a curve of effective windows across a population.

She shows how a standard can be gamed by targeting the middle while abandoning the edges. If you set the minimum window based on a typical reaction, you can still leave a large fraction of people with windows too narrow to use. The system will look fair on average while being cruel in practice.

So the speaker proposes a second layer of compliance. Not only minimum window, but coverage. A compliant design must provide effective windows for a large fraction of real users, and it must provide reversibility for those outside the typical band.

The room is uneasy now, because complexity has returned. The simple mark is becoming a real promise, which means it is becoming harder.

This is the moment when someone asks the obvious question, the question that always arrives when standards become complicated.

Who pays for this.

The speaker does not answer with moral slogans. She answers with the simplest economic observation in the world. Someone always pays. The only question is whether the cost is paid upstream as protection, or downstream as damage.

Delay has always been subsidized by the people who suffer its consequences.

So she proposes that cost be internalized. If a company wants to gate action, it must pay for stronger reversibility. If a company wants to keep warnings gentle, it must pay for longer effective windows. If a company wants to flood the channel, it must pay for signal integrity penalties.

The room begins to feel the shape of the future, and it is not a gadget future. It is a governance future built out of timing receipts and test suites and measurable levers.

Then another fear rises, quieter than capture, but deeper.

If you create timing receipts for everything, you create a record. Even if the record is minimal, even if it is a diagram, records have a way of being repurposed.

A man speaks about privacy. He speaks about how diagrams become profiles. How profiles become passports. How passports become gates.

The speaker nods again, and she agrees that the danger is real. She says that the answer is not to avoid measurement, because measurement already exists. The answer is to constrain what measurement can be used for.

She proposes that timing receipts be legally limited to accountability for the interface, not surveillance of the person. She proposes that receipts be anonymized by default. She proposes that a person be able to summon a receipt for a specific event without giving away their entire life history. She proposes that the receipt expire, domain by domain, so that a moment of delayed access does not become a permanent label.

The audience listens, and they realize that something has happened in the hearing that is rare. The discussion has moved from blaming a person to designing a world.

But the speaker knows the last trick, the trick that will haunt the entire project.

The trick is that prediction can be used to seduce people into accepting lower access. You can show them a beautiful forecast and call it empowerment. You can give them high clarity while shrinking their levers.

So she says, slowly, that the certification mark must include a second mark, one that separates legibility from control.

Forecasting compliant is not the same as access compliant.

The room goes very still because that sentence changes the meaning of modern life. It implies that many things people have been calling progress are merely improved prediction layered on top of shrinking power.

She says that in a volumetric universe, the most dangerous kind of comfort is readable helplessness.

Then she ends the session with a reminder that sounds almost gentle, but is not gentle at all.

A standard is not a promise unless it can survive adversarial design.

A mark is not a guarantee unless it can resist capture.

A lever is not a lever unless the world moves with it in time.

The timeline remains on the screen when people leave. It remains because it is the most honest object in the room.

Not a law. Not a speech.

A line with dots, and a shaded band, and the brutal calm of seconds.

The room empties the way a theater empties after a show that nobody is sure they enjoyed. People file out in small clusters, speaking softly, already rewriting their opinions in real time. Outside, the city continues to move, lights change, buses hiss, a siren passes somewhere far away, but something has shifted, not in the streets, in the way the streets will be judged.

The timeline stays on the screen after everyone leaves. It stays because nobody wants to be the one to turn it off. Turning it off would feel like denial. Turning it off would feel like choosing to go back to the old moral reflex, the reflex that points at the nearest human and calls the problem solved.

A technician waits until the last chair is pushed in, then walks to the console. He hesitates, then instead of shutting the display down, he saves the diagram, not as an image, as a record of timing, a record of gaps, a record of which levers were real and which were ornamental. The new object of public argument is not a speech, it is a sequence of seconds.

Within days, the headline is not about the commuter train. The headline is about the window.

People begin to repeat the phrase effective window the way people once repeated the phrase due process. They say it at dinner tables. They say it in interviews. They say it in arguments about credit, about medicine, about housing, about safety, about the small humiliations of modern life where a person is told they had a choice while being offered only a button that records compliance.

In the weeks that follow, prototypes appear. They do not look glamorous. They look almost boring, which is how serious inventions often begin.

A new kind of receipt is attached to consequential prompts. It lists when the signal arrived, when the lever became available, when reversal ended, and how much time was left in the effective window. It is short. It is plain. It does not try to persuade. It does not try to soothe. It simply tells the truth about timing.

Some people hate it immediately. They say it will cause anxiety. They say it will overwhelm users. They say it will slow everything down. They say that life is already complicated enough without reminding everyone that every decision has a clock.

They are partly right. It does cause anxiety. It does overwhelm users. But it also reveals something that has been hiding in plain sight.

Many of the most stressful moments in modern life are not stressful because the future is unknown. They are stressful because the future is known too late, and the lever arrives after the branch has closed. The receipt does not create the pain. The receipt names it.

Companies respond the way companies always respond to a new constraint. First they promise alignment. Then they search for loopholes. Then they hire people whose job is to shape the definition of compliance until compliance becomes cheap.

Early failures arrive fast.

A payment app adds an undo feature, but the undo is buried behind several screens, and the effective window expires before a typical person can reach it. The receipt reveals the trick in seconds. The app is compliant on paper, but not in practice.

A medical consent interface adds a plain language summary, but the summary is still too long, still too vague, still too polite. The words are readable, but the meaning arrives late. The receipt reveals the trick again. The consent exists, but the window is still too narrow.

A transport warning arrives earlier, but it is delivered as a gentle suggestion rather than a directive, and the lever remains gated. The receipt reveals the trick again. Early signal without early action is not protection. It is decoration.

These failures are public now. They become case studies. They become memes. They become training material. A new kind of literacy begins to spread, not technical literacy, timing literacy. People start asking a simple question before they click. They start asking whether the lever is real. That question changes more than any slogan ever could.

The certification mark evolves. It stops being a single seal. It becomes a layered label, a small stack of facts. Verified window. Verified lever coupling. Verified reversal. Verified signal integrity. Verified timing receipt.

The mark does not guarantee goodness. Nothing guarantees goodness. The mark guarantees only this, that if a person is harmed, the clock will not be allowed to hide.

In that sense, the mark is not a shield, it is exposure.

Resistance intensifies precisely because exposure works.

There are hearings, not like the first one, more procedural, more technical, more full of people who speak in a way that makes their motives sound like math. There are arguments about domain thresholds, about how many seconds count as enough, about whether a typical nervous system is an acceptable benchmark, about what to do with those who sit outside the typical band. There are arguments about how to keep the receipts minimal, how to keep them from becoming a new passport, how to stop timing diagrams from turning into identity.

All of these arguments are real. None of them are trivial.

Because there is a deep tension at the center of the new world.

If you measure access, you can defend people from unfair blame. You can show that the lever arrived too late. You can show that clarity was withheld. You can show that the option was ornamental. You can shift accountability upstream to those who designed the window.

But if you measure access, you can also classify people. You can label them. You can decide they are too slow, too noisy, too inconsistent. You can decide that the safest choice is to restrict what they can do, not because you hate them, because you believe you are helping them.

This is how care becomes a cage.

The only way out is the principle that the panel introduced and that the public repeats with growing intensity, protection through reversibility, not removal.

That principle begins to show up in places nobody expected.

Contracts become less absolute. Cooling periods expand. Undo becomes standard for more classes of action. Appeals are forced upstream, into windows that still matter. Decisions that used to be instant become deliberately slow, not slow in the old way that steals agency, slow in a new way that preserves it.

A bank transfer above a threshold waits. Not for punishment. For reversal.

A public safety alert arrives with a lever that does not require permission. Not because humans are trusted, because physics is indifferent.

A housing denial must include a list of actionable factors that can be changed inside a real window, and if nothing can be changed inside the window, the denial cannot be framed as a choice. It must be framed as a gate, and gates must be justified differently than choices.

This is the quiet revolution of timing. It does not feel heroic. It feels like friction added in the right places. It feels like a world that stops pretending.

Then the counter revolution arrives, also quietly.

Some institutions begin to treat reversibility as a threat. Reversibility keeps branches open, and branches open means unpredictability. Unpredictability is expensive. Unpredictability makes forecasts less clean. Unpredictability makes planning harder. Unpredictability makes people harder to manage.

So the pressure begins to return, the pressure to close branches earlier, to tighten the effective window, to reduce reversal, to push decisions into instant outcomes again.

The argument is always the same. Stability. Safety. Efficiency. Improvement.

And because these words sound good, they work, unless the receipts exist. Unless the test suites remain public. Unless the certifiers remain multiple. Unless the standards remain auditable. Unless capture fails.

The public learns an old lesson in a new form.

A standard is only as strong as its resistance to capture.

That is why the most important part of the new world is not the seal. It is the adversarial test.

A device that can pass a friendly test can still steal your window. A device that can pass an adversarial test is forced to be honest about timing.

A new kind of journalist appears, part engineer, part ethicist, part nuisance. They test interfaces the way crash testers test cars. They try to fail them. They measure how long it takes to find undo. They measure how long it takes to reach a human when the stakes are high. They measure how many prompts a person must endure before a critical signal can be recognized. They measure the difference between available and usable.

This is where the moral landscape changes permanently.

Responsibility used to be assigned with stories. You were warned. You agreed. You should have known. You clicked.

Now responsibility is assigned with diagrams. Now the clock is evidence.

The public begins to grasp the simplest truth of the Volumetric Time Model as it applies to ethics.

What matters is not only what was true. What matters is what was reachable in time.

Every time you are offered a choice, you are being offered a branch. If the branch arrives late, the branch is fake. If the lever does not couple, the branch is fake. If the appeal lives downstream from the last meaningful split, the appeal is fake.

The future becomes a place full of fake branches unless someone defends the windows.

Even if the world builds perfect timing standards, even if every interface becomes honest, even if every lever is real, the volume is still finite. The cone of access is still narrow. Biology still has latency. Attention still has limits. Distance still filters signals. Noise still exists.

No certification mark can make you a god.

All it can do is stop others from pretending you were a god when they designed your windows to fail.

That is the difference between justice and theater.

As night falls, the first hearing is already becoming a story people tell about the moment timing became visible. The actual victims fade into the background, as victims often do. The timeline remains the main character. The shaded band remains the plot twist. The effective window remains the new moral unit.

And the speaker, the one who introduced the phrase clock owners, is interviewed again. This time she is asked whether the new standards will fix everything.

She answers with the calmness of a person who understands geometry.

No standard will fix everything. The volume will still be there. The future will still be large. Access will still be limited.

But the standards can do one crucial thing. They can stop the modern trick of giving people prediction and calling it power.

They can stop the modern trick of giving people a button and calling it choice.

They can make the levers honest.

They can force accountability upstream, toward those who allocate delay budgets and design the shape of reachable futures.

FORMULA DECK (Chapter 10)

Responsibility Tracks Access, Receipts, Windows, and Standards

1. Events, outcomes, and the region. Let spacetime be represented as (M, g). Fix a compact region V ⊂ M with boundary ∂V, and let H be the admissible history set on V. Consider a consequential event specification e inside V. For each admissible history ϕ in H, let O(e, ϕ) denote the outcome associated with event e under history ϕ. This supports comparing the same event specification across different globally consistent histories.
2. Access as a growing record over time. An embedded observer does not access the full history ϕ directly. Instead the observer receives an accumulating record of accessible data. Let D_t denote the data accessible up to time t, and define the information set F_t as the sigma algebra generated by that record, written F_t = σ(D_t). As time increases, the accessible record grows. For s ≤ t, we have F_s ⊆ F_t. This models the idea that the admissible set H can be fixed while the observer access continues to update.
3. Lever availability and operational access state. Many consequential outcomes can only be altered if a relevant lever is available. Let L_t(e) be a binary indicator taking values 0 or 1. L_t(e) = 1 means the relevant lever for event e is

available at time t, and L_t(e) = 0 means it is not. Define the observer operational access state as A_t(e) = (F_t, L_t(e)). This combines what the observer can know at time t with what the observer can actually do at time t.

4. Branch close time as the timing of irreversibility. For each event e, define a branch close time, written t_b(e). This is the last time at which an intervention could still alter the outcome O(e, ϕ). After t_b(e), the relevant branch has closed and the outcome is operationally irreversible for that event. This is the formal location where too late becomes a real statement.
5. Effective window length for real choice. Let t_s(e) be the time a decisive signal becomes accessible to the observer for event e, meaning the earliest time the signal is present in D_t. Let t_l(e) be the time the relevant lever becomes available, meaning the earliest time when L_t(e) = 1. Define the effective window length as follows. The effective window length for event e equals t_b(e) minus the later of t_s(e) and t_l(e). If this quantity is positive, the observer had a usable window in which the signal and the lever overlapped before irreversibility. If this quantity is zero or negative, then the lever is operationally absent for that event, even if it exists in principle.
6. Reversibility window and undo horizon. Some outcomes can be reversed for a limited time. Let t_r(e) be the end time of a reversal or undo

period for event e. Define the reversibility window length as follows. The reversibility window length for event e equals t_r(e) minus the later of t_s(e) and t_l(e). A practical safety requirement is domain dependent. The reversibility window length should exceed a minimum value, written Δt_min(domain(e)), when stakes are high. This encodes protection through reversibility rather than removal of options.

7. Signal integrity as timely detectability. A warning can arrive and still be unusable if it cannot be detected and understood in time. Let T_det(e) denote the time at which the observer correctly recognizes the critical signal for event e. A minimal signal integrity requirement is that timely detection is likely. In practice, the probability that T_det(e) occurs within the effective window should be at least a chosen level p_0. This formalizes the idea that flooding the observer with background prompts can shrink usable access even when signals technically arrive early.
8. Lever reality as outcome coupling, not interface motion. A lever is real only if action through it measurably changes outcomes. Let U_t(e) denote the observer action applied through the lever at time t when L_t(e) = 1. Define a steering coupling measure that asks whether actions U_t(e) influence the future outcome of event e

after some delay, and test this against null controls. If the measured coupling is indistinguishable from a null baseline, then the displayed lever is ornamental. It moves the interface but not the world.

9. Timing receipt as a minimal audit object. Define a timing receipt for event e as the tuple $R(e) = (t\_s(e), t\_l(e), t\_b(e), t\_r(e))$. From this receipt one can compute the effective window length and the reversibility window length without storing content. The receipt is meant to be a minimal record that makes timing accountable while limiting exposure of personal details.
10. Responsibility as a function of operational accessibility. The core claim is that responsibility should be calibrated to what was operationally accessible. In practice, responsibility weight is high when three conditions hold. The effective window length is positive. The signal is detectable within the effective window with probability at least p_0. The lever has real outcome coupling above null. Responsibility weight collapses when any of these conditions fails. This shifts accountability away from downstream blame and toward upstream control of windows, levers, and signal integrity.
11. Standards and adversarial testing. Let S be a public test suite that samples representative events e within a domain and measures effective window length, reversibility window length,

detectability, and lever coupling. A minimal compliance requirement is that across S, effective windows meet domain minimums, reversibility meets domain safety thresholds, detection probability is above p_0, and lever coupling exceeds null baselines. This formalizes the chapter conclusion that fairness depends on auditable windows and real levers, not on late acknowledgements.

Effective and Reversible Windows

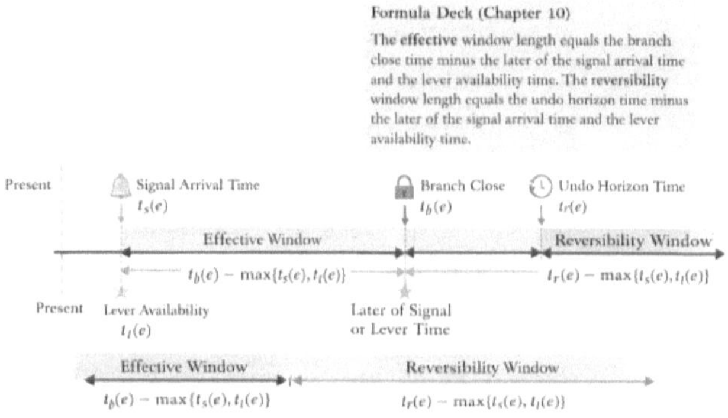

Chapter 11

If VTM is Real

Imagine a museum built for a civilization that has outlived the river metaphor.

There is no long blue ribbon painted on the wall. There are no arrows, no flowing lines, no glowing streams to suggest that time moves like water and carries you along. Instead there is a dark room and a single object floating in the center under soft light.

It is a sphere.

Not a globe of Earth, not a planet, not a star. A sphere that represents a life.

A guide stands beside it, calm, confident, almost cheerful in the way physicists can be when they are about to ruin a comforting idea. The guide does not speak about destiny or fate. The guide speaks about geometry.

You were born inside this sphere. You spend your entire life moving through it. The sphere is finite. It has a boundary. You do not reach the boundary all at once. You reach it one step at a time, one decision at a time, one breath at a time. You cannot hold the sphere in your hands, you can only traverse it.

A child raises a hand, because children still do that in the future, and asks where the future is located.

The guide points, not forward, not upward, but outward, toward the surface.

The future is all around you. The past is behind you. The present is the moving point where you stand. The boundary is where traversal ends.

The child looks confused, because the child has been trained by stories to imagine time as a hallway and life as a line. So the guide does what good communicators have always done, from Newton to Einstein to the great popularizers who made modern physics feel like a shared adventure.

The guide gives the child a metaphor that can survive contact with reality.

Life is not a line on a page. Life is a path in a room. You do not turn pages, you walk. You do not flip ahead to see the ending, you move toward it, and the room reveals itself only as fast as your senses can take it in.

The child nods a little, then asks the next question, the one adults ask too, just with better vocabulary.

If the sphere is already there, does that mean everything is fixed.

The guide smiles, because the question is natural and the answer is not as simple as people fear. The guide points again, this time to a cluster of faint threads inside the sphere, like hairline filaments in glass. Each thread is a possible lived history. They overlap. They twist. They merge. They split. Some are thick, some are thin, some end early, some stretch toward the boundary.

The guide says that the sphere contains admissible histories. Many paths exist. Not all paths are reachable from where you stand. Not all paths remain reachable as you move. What you call choice is the act of selecting a direction through this dense forest of trajectories.

Then the guide says the sentence that changes the room.

Existence does not guarantee access.

The child frowns again, because the sentence sounds like a warning sign and children are sensitive to warning signs. The guide explains with a simple image.

You stand inside the sphere with a small flashlight. You cannot light up the whole sphere. You cannot see every thread. You cannot know the full structure. Your flashlight has a cone. The cone sweeps across nearby regions. It reveals some paths. It hides others. It does not lie, it simply cannot illuminate everything.

The cone is your access.

The child asks what determines the cone.

The guide lists the same factors that determine so much of human life, distance, noise, attention, delay, the speed at which signals travel, the time it takes for minds and bodies to recognize meaning, the time it takes for actions to couple to consequences. The guide says that even if the sphere is real, even if every admissible path exists as a thread within it, you only live inside what becomes reachable through your cone of access.

This is the deepest shift that the Volumetric Time Model forces on the reader. The central mystery of human life is not whether reality exists. The central mystery is what can be reached in time.

A second display lights up in the museum. It shows a familiar modern scene, a person standing at a crosswalk looking at a phone. A message arrives. A warning arrives. A prompt arrives. A button appears. The person taps the button. The person feels like they acted.

Then the display overlays a second layer, invisible in daily life but obvious once you learn to look for it. A shaded band appears around the moment the prompt arrives. It represents the effective window, the interval during which a lever, if real, could still steer the outcome.

The shaded band is narrow. Sometimes it is so narrow it is almost a line. Sometimes it is wide enough to include a conversation, a reconsideration, a reversal. Most of the time it is narrower than you want to admit.

The guide turns back to the sphere and points to a region near the child, a region filled with branching threads. The guide explains that these are the moments when your life has real branching in the sense that matters. The moments when your motion can still couple to different outcomes. The moments when a lever is not decoration.

The guide also points to regions where the threads are almost parallel, where the paths are tightly bundled and the branching is minimal. In those regions you may still receive information. You may even receive accurate information. You may feel clarity. But clarity can arrive after branching has already collapsed. Clarity can arrive after coupling is gone.

This is the first hard lesson of the ending of this book, prediction is not power.

The room in the museum is quiet, because the metaphor has done its job. People can feel the difference between being told a future and being able to steer a future. They can feel why late certainty can feel like a trap. They can feel why modern life can produce a strange kind of despair that does not come from ignorance. It comes from legibility without leverage.

The guide does not say those phrases as slogans. The guide keeps it simple.

You can be right about what will happen and still be unable to change it. You can be surrounded by truth and still be unable to use it. You can be present in a moment and still be locked out of the outcomes that moment determines.

Existence does not guarantee access.

The child looks back at the sphere again. The child asks the question that finally matters.

If the cone is small, how do I live.

The guide does not answer with philosophy. The guide answers with practice.

You protect the cone. You protect the conditions that keep the cone honest. You protect the places where feedback is fast and clear. You protect reversibility. You protect signal integrity. You choose lives where the world still answers when you touch it.

The guide says that this is not self help. This is not a motivational poster. This is physics translated into ethics.

Then the guide walks the group to the far wall, where the museum has built a second exhibit for adults.

It is a simple panel titled The Great Confusion.

On the panel are two statements that most people treat as the same.

Knowing.

Steering.

The panel explains that modern technology has become excellent at knowing. It can forecast. It can score. It can predict. It can label risks and likely outcomes with increasing precision. It can illuminate parts of the sphere with bright cold light.

But steering depends on something else, on coupling. On whether a lever exists. On whether the lever arrives inside the effective window. On whether the feedback loop is short enough for action to matter.

The panel shows a cruel curve. As delays increase and windows shrink, steering becomes harder. Yet prediction can remain strong, sometimes it can even become stronger, because the reachable set of futures narrows and the world becomes easier to guess.

The guide does not call this curve a paradox in this room. The guide calls it a warning.

When the future starts to feel obvious, check your levers.

The child reads the sentence and laughs a little, because it sounds like a joke. The adults do not laugh, because they have lived the sentence.

The guide brings the group back to the sphere one last time and asks them to notice something simple.

Most of the threads inside the sphere are never seen by any person. Not because the threads are not real. Because the cone is limited. Life is not lived in the full space of admissible histories. Life is lived in the thin region where access meets action.

That region is what you can protect.

This is why the ending of this book is not a debate about whether the universe is fixed. The ending is about what it means to be a human agent inside a finite volume. It is about the difference between being informed and being empowered. It is about how to build a world that does not confuse warning signs with choices.

The guide dims the light in the room. The sphere becomes darker. Only a small moving point remains visible, the path traced by one life. It moves, slowly, steadily, toward the boundary.

The point is not heroic. The point does not glow with destiny. It simply moves, and as it moves it illuminates a narrow cone ahead, enough to choose, enough to care, enough to steer in the places where steering is still possible.

The guide ends the tour with one last image.

You do not get to own the sphere. You do not get to rewrite the volume. You do not get to see every thread.

You get a cone of access.

Your task is to make sure the cone is real.

Your task is to guard the places where the world still answers.

The Tome Of Access

The sphere in the museum is a useful lie. It is useful because it makes the idea visible. It is a lie because no one ever stands still inside life. The sphere does not sit in a room waiting for you to observe it. You move, you age, you get tired, you get distracted, you get hungry, you fall in love, you lose people, you stare at the ceiling at night and negotiate with your own thoughts. You do all of this while the point that represents you keeps traveling through the volume.

Your access is not a property of the universe alone. Your access is a property of the universe plus your position plus your limits plus the rules that determine which levers appear and when they appear.

That is why the cone matters more than the sphere.

The cone is the difference between the existence of truth and the arrival of truth. The cone is the difference between a signal that is physically present in the world and a signal that reaches you in time to change your path. The cone is the difference between knowing and steering.

It is tempting to imagine the cone as a spotlight you control. It feels good to imagine that. It flatters the mind. It makes the self feel like a captain.

In reality the cone behaves more like a weather pattern. It has shape. It has drift. It has limits. It has blind spots. You can influence it, but you do not own it.

The first limit is simple physics. Signals take time to travel. Light from the Sun takes minutes to arrive. Light from distant galaxies takes millions of years. The night sky is a reminder that access is always delayed. You never see the universe as it is now. You see the universe as it was when the photons left their source. Even the face of a person across a room is a fraction of a second in the past by the time it reaches your eyes.

This is not poetic. It is a physical constraint. It is the baseline condition of existence.

If you accept that, a strange calm arrives. You stop demanding impossible immediacy from reality. You stop treating delay as a personal insult from the cosmos. The world is delayed because the world is extended in space and signals have a finite speed.

Then the second limit arrives, and it hurts more, because it is not the cosmos, it is you.

A signal can arrive at your senses and still fail to become usable. You might not notice it. You might notice it but not recognize it. You might recognize it but not understand it. You might understand it but not act in time. You might act and still not couple to the outcome because the lever arrives too late or because the world answers too slowly.

This chain is long. Each link is a place where access can collapse.

Think of a simple moment. You are walking, a car turns too fast, you hear the engine, you see the motion, your body shifts. In that moment your cone is wide and fast. Your senses and muscles are built for this. The loop from signal to action is short. The world answers quickly. You step back and the car misses you. You do not need philosophy. You need latency and reflex and a lever that is your own legs.

Now think of a different moment. You receive a message. The message is true. The message matters. But your mind is already saturated. You have opened ten other prompts. You have been trained to tap away banners. Your finger moves before your meaning arrives. You register receipt. The signal enters your life as noise. The loop is broken. The lever is ornamental. The branch point closes and you learn later that you were late.

In both cases the universe is the same. In both cases the truth existed. The difference is the shape of your cone.

This is why the Volumetric Time Model is not a cold deterministic story. It is a story about what reaches you, and how it reaches you, and whether it arrives in time to matter.

The third limit is the most modern limit. It is not physics and it is not biology. It is design.

Many of the levers in modern life are not located in your body. They are located behind prompts, gates, confirmations, and delayed permissions. They are located inside interfaces. They are located inside policies. They are located inside schedules that you did not set.

These levers change the cone in two ways.

They change the cone by controlling what information arrives and when it arrives. They also change the cone by controlling which actions are allowed and when they are allowed.

If you want to understand why people feel powerless in a world full of information, look here. Do not look for villains. Look for delayed levers.

A person can be surrounded by forecasts and still have no steering. A person can be given perfect warnings and still be trapped because the lever arrives after the branch point. A person can be praised for engagement and still be denied a real option that couples to the outcome.

This is why existence does not guarantee access. It is not only a cosmic statement. It is a civilizational statement.

The cone can be narrowed on purpose without anyone ever admitting they narrowed it. It can be narrowed by adding one extra step. It can be narrowed by moving the lever deeper into menus. It can be narrowed by offering reversibility that expires too soon to reach. It can be narrowed by flooding the channel until signal integrity collapses.

When this happens the person is still told they have choices. They are still asked to confirm. They are still asked to acknowledge. They are still treated as responsible.

But the geometry has changed. The reachable set of futures has been reduced.

This is where the emotional experience of fate enters daily life. Fate is what a human being calls it when the cone becomes narrow and the world remains highly legible.

It is important to say that again because it is one of the strangest psychological shifts of the modern age.

People do not feel trapped because they are ignorant. People often feel trapped because they are informed too late. They see the cliff, they see the diagnosis, they see the market move, they see the relationship decay, they see the weather forming. They see it clearly. They see it in high resolution. They see it with charts and notifications and predictions. They see it while the levers that could have changed the path are either hidden, delayed, or already gone.

That combination produces a specific kind of despair. It is not the despair of darkness. It is the despair of bright light with locked doors.

The cone also has a cost. Every widening of access requires energy. A telescope gathers faint photons. A sensor requires power. A mind requires calories. Attention is not free. Attention is a budget.

This matters because it prevents a naive fantasy. The fantasy says that we can fix the problem by giving everyone more information. The fantasy says we can widen the cone by pouring more data into it.

In practice more data often narrows the cone because it consumes the budget that turns signals into meaning. A channel can be filled until nothing gets through. A mind can be loaded until every prompt becomes the same blur. A community can be saturated until urgency loses its shape.

So the cone is not simply a quantity. It is also a quality. It depends on signal integrity. It depends on clarity. It depends on the difference between one meaningful warning and a thousand polite reminders.

The deepest point here is that access is not the same as coverage. You can cover a person in signals and still deprive them of access. You can bury meaning under volume.

That is why the future does not need to be imagined as a world with more screens. It can be imagined as a world with more honest timing. A world where signals are fewer and sharper. A world where levers are real and early. A world where reversibility exists long enough to reach. A world where the cone is not widened by noise but by integrity.

If this sounds like ethics, it is ethics. But it is ethics that behaves like engineering. It is ethics that can be measured in seconds and in the placement of levers and in the shape of effective windows.

Now return to the museum sphere. Imagine the threads again. Some regions are densely branched. Some regions are tightly bundled. As you move, the shape of the reachable set changes.

Here is a fact that seems obvious once said, and that is why it is powerful.

You only experience choice where your cone overlaps with branching.

If branching exists but the cone does not reach it in time, then the branch might as well not exist for you. The path may be admissible in the full volume, but it is not reachable from your lived position. It is outside your operational horizon.

This is why two people can live in the same physical world and inhabit radically different moral worlds. Their cones are different. Their windows are different. Their levers are different. One person lives in a region of reachable reversal. Another person lives in a region of irreversible gates. One person lives in a region where feedback is fast and clear. Another person lives in a region where feedback arrives as a receipt.

The Volumetric Time Model makes this difference visible. It turns inequality into geometry.

It also reframes a quiet truth about modern culture. A great deal of our public anger is really a dispute about cones. People are not only fighting about values. They are fighting about what is reachable. They are fighting about who gets early warnings and who gets late warnings. They are fighting about who gets real levers and who gets ornamental control.

Once you see that, you start to understand why prediction can feel like oppression when it arrives without steering. You start to understand why people can be exhausted by guidance. You start to understand why some promises feel like insults. You start to understand why late clarity can feel like a trap.

This brings us to the next step of the ending. The strange curve that appears when steering collapses. The moment when the future becomes easier to guess precisely because fewer futures remain reachable.

When the future starts to feel obvious, check your levers.

The Prediction Paradox

There is a kind of clarity that arrives late.

It arrives with the calm confidence of a screen that has already made up its mind. It arrives in neat sentences and clean charts and gentle signals that feel almost compassionate. It arrives when the future is no longer a wide landscape of possibilities, it is a narrow corridor with the lights already pointing in one direction.

Most people assume that clarity is a gift. They assume that seeing farther is the same as steering farther. They assume that prediction is a form of power.

Then they live long enough to learn the opposite.

The moment the future becomes most obvious can also be the moment it becomes least reachable.

That is the paradox, and it is not philosophical, it is geometric.

A storm forms over the ocean. In an earlier age a farmer watched the sky, smelled the air, listened to the birds, and if the signs were strong he pulled tools inside, he tied down loose wood, he moved animals, he changed his path through the day. The warning was crude, but the lever was immediate. The loop from signal to action was short.

Now imagine the same storm in the present world.

The storm is tracked by satellites, measured by sensors, modeled by machines that drink data by the ocean. The forecast is precise. The landfall probability is mapped down to the neighborhood. The timing is plotted with beautiful confidence. The future becomes legible with a sharpness that would have felt like magic to past centuries.

Then the first evacuation notice arrives.

It arrives after the highways are already clogged. It arrives after fuel has already disappeared from local stations. It arrives after hotels are already full, after flights are already canceled, after the realistic options have already collapsed into a smaller set.

The forecast is perfect. The steering is weak.

The person is told what will happen, then is asked to accept what will happen.

This is not because people have become lazy or irrational. This is because the branching has already narrowed by the time the clarity becomes widely accessible. The cone arrives when the branch points are closing. The information is true, but it is late in the only way that matters.

The future is readable, but the levers are already far away.

A second example is quieter and more personal.

A doctor sees a trend in a patients data. A risk curve rises. A warning threshold is crossed. The future becomes visible. The likely outcome becomes obvious. A screen lights up with a suggestion to act.

The patient receives a calm notification, a gentle instruction to schedule an appointment, a polite note that early intervention improves outcomes. The words are correct. The words are responsible. The words are also slow.

By the time the appointment exists, the branch point might already be behind the patient. Not because medicine is useless, because time is indifferent. Some processes accelerate. Some curves steepen. Some outcomes lock in quickly once the last small window closes.

Again the paradox appears. The better the prediction, the more the person can feel trapped. The mind experiences a strange mix of enlightenment and helplessness, as if the universe is finally telling the truth, but only after the moment when truth could be used.

This is why modern fear often feels different from old fear. Old fear came from darkness. Modern fear can come from bright light.

A third example is the one nobody wants to admit, because it makes people feel guilty.

A relationship begins to decay. Not with a single betrayal, not with a cinematic blow up, but with tiny delays; tiny failures of response, small absences that slowly accumulate into a new shape. The future becomes visible long before the end. People see it in the way messages take longer to answer, in the way laughter changes, in the way attention shifts elsewhere.

Clarity arrives. The ending becomes easy to guess.

And yet the levers that could have changed the outcome were earlier, smaller, almost childish in their simplicity. A conversation that happened sooner. A repair attempt made before resentment hardened. A choice to protect a short feedback loop rather than letting the loop stretch until it snapped.

When the ending becomes obvious, it can already be late.

This is the lived shape of the paradox.

As time passes and as branching narrows, prediction can become easier.

Not because the universe is helping you. Because fewer futures remain reachable.

Think of a wide river near its source, braided into many channels, spilling across a plain. From above it looks chaotic. Water can go here or there. It can split and rejoin. It can carve new paths after rain. It is hard to predict the exact line.

Now imagine the same river as it approaches the sea through a narrow canyon. The channels merge. The choices collapse. The water has fewer options. The path becomes obvious.

Prediction becomes easier because the degrees of freedom are gone.

A life inside a finite volume behaves the same way. When branching is rich and the reachable set is wide, prediction can be hard. Small actions matter. Tiny disturbances can push the trajectory into a different neighborhood of futures. The world looks noisy, uncertain, full of possibility.

Then as certain windows close, as reversibility decreases, as options narrow, the set of reachable futures becomes smaller. The future becomes more predictable, not because it is being revealed, but because it is being reduced.

This is the moment when certainty becomes a warning signal.

Late certainty is often the signature of a closed corridor.

It is not the universe giving you a gift. It is the universe telling you that the map has fewer roads.

The paradox shows up in markets with brutal clarity.

A market can look wildly unpredictable when many participants can still act, when feedback is fast, when information is messy, when prices can still be steered by millions of small choices. Then as a crisis approaches and leverage collapses and options narrow, the future can become obvious. People begin to see the cliff. They begin to speak in the same phrases, they begin to move in the same direction, they begin to react in a synchronized way.

The crash feels inevitable in retrospect. In the final days it can feel inevitable in advance.

But in those final days the ability to steer is already weak. The person can predict the move, but cannot escape it because the exits are full of bodies and the locks are already turning.

This is forecasting without power in daily language. You can see it coming. You cannot change it. You may not even be able to avoid it.

And the cruel twist is that the clarity itself can make people feel more trapped. The mind interprets prediction as a promise that action should be possible. When action fails, the mind reads the failure as personal.

The mind says I knew, why did I not stop it.

The correct answer is that knowledge and steering are not the same coupling. One is about information. The other is about influence. They can separate. They often separate. They can separate cleanly.

A useful way to feel this difference is to imagine a person watching a distant avalanche. The person has a perfect view. The person can predict where the snow will go. The person can even estimate when the sound will arrive. The person can be correct about everything.

The person still cannot stop the avalanche.

Prediction is not force. Prediction is not leverage. Prediction is a view.

In a finite volume, the cone of access can widen in the sense of knowledge while narrowing in the sense of reachable steering.

This is why the future can become readable while becoming unreachable.

There is another twist that makes the paradox more dangerous.

Prediction can be used as sedation.

If you can show people a beautiful forecast, you can make them feel included. You can make them feel informed. You can make them feel like participants in the story of what will happen.

Meanwhile you can quietly remove levers. You can delay action, gate action, bury reversibility, and tighten effective windows.

You can offer legibility as a substitute for control.

This is the most modern kind of trap. The trap does not hide the future. The trap displays the future. The trap says look, we are being transparent, we are showing you the truth.

Then the trap asks for acknowledgement.

It thanks you for your feedback.

It proceeds.

In this environment a new skill becomes essential.

Not the skill of predicting. Many tools can predict now. Not the skill of collecting more information. That often adds noise.

The essential skill is to recognize when certainty arrives too late.

When the future starts to feel obvious, check your levers.

This is not a cynical slogan. It is a practical diagnostic of where you stand inside the volume.

If the future has narrowed into a corridor, the first question is not why is this happening. The first question is where did the branch points go. The second question is what windows closed. The third question is whether there is any reversibility left.

Often the only honest leverage in a late corridor is not to fight the forecast. It is to move upstream in time.

That does not mean time travel. It means building habits and institutions that keep windows open earlier.

It means making decisions reversible before they become irreversible. It means protecting short feedback loops. It means resisting the temptation to treat a late warning as proof that the world is fair. It means recognizing that a warning delivered after the only effective window is a receipt, not a lever.

This is the hidden advantage of understanding the paradox. Once you see that clarity can arrive after control is gone, you stop worshiping clarity as a moral object. You stop confusing being informed with being empowered.

You begin to demand a different kind of promise, not a promise of perfect prediction, a promise of real levers that exist in time.

There is also a personal consequence that is difficult but liberating.

Many regrets are built out of the paradox. People replay a moment and say it was obvious, I should have known, I saw the signs. They treat the clarity of hindsight as evidence of failure.

Sometimes that is true. Often it is not. Often the clarity arrived because the corridor had already formed. The future became obvious because options had already narrowed. The regret is real, but it does not prove that a different outcome was reachable at the moment the mind claims it should have been.

This does not absolve a person of all responsibility. It places responsibility where it belongs, on the interaction between access and branching. It reminds you that a life is not judged by what could have happened in the full volume. A life is judged by what could be reached from the actual position you inhabited with the cone you had.

The paradox therefore becomes an ethical test.

If a society becomes better at prediction while becoming worse at preserving reach, it will produce a population that feels haunted by foreknowledge. People will feel that they are watching their own lives in advance while being unable to steer. They will feel as if the universe is narrating their future with increasing accuracy, and they will confuse narration for freedom.

If a society becomes better at preserving effective windows, reversibility, and signal integrity, it will produce a population that may see less, but can do more. The future will feel less obvious, because more futures remain reachable. Uncertainty will increase, but it will be the uncertainty of real branching, not the uncertainty of darkness.

This is a difficult trade for modern minds, because modern minds have been trained to crave legibility. They have been trained to treat uncertainty as ignorance. They have been trained to treat prediction as progress.

The Volumetric Time Model suggests a different measurement of progress.

Progress is not the ability to describe what will happen.

Progress is the ability to keep more of what could happen reachable.

When the future starts to feel obvious, check your levers, and then move upstream, because the only reliable answer to late certainty is earlier coupling.

If prediction is a view, then agency is a grip. It is the difference between watching a map and moving a steering wheel. It is the difference between recognizing the cliff and having a real lever before the edge.

The easiest mistake in modern life is to treat agency as a feeling. You wake up energized and you think you have agency. You wake up tired and you think it is gone. You hear a confident forecast and you feel included. You hear uncertainty and you feel powerless. But the Volumetric Time Model does not care about mood. It describes agency as a physical relationship between action and consequence, a coupling that can be measured by whether the world answers your motion in time.

Agency is preserved when the coupling remains above threshold. That sounds abstract until you name the ingredients that keep coupling alive.

Fast feedback.

Clear meaning.

Real levers.

Reversibility when the stakes are high.

Signal integrity, so urgency is not drowned by noise.

These are not inspirational phrases. They are engineering constraints for a human life.

Start with the simplest kind of agency, the kind your body was built for. You touch a hot surface, you pull your hand away. You do not write a plan. You do not consult a chart. The loop is short. The meaning is clear. The lever is your muscles. The world answers instantly. This is agency in its original form, a tight circuit between perception and motion.

Now compare it to the kind of agency that fills modern days. You get an alert about a payment. You get a notification about a health marker. You get a warning about a risk. The loop is long. The meaning is wrapped in polite language. The lever is hidden behind steps. The world answers late. The person feels like they acted because they acknowledged, but the outcome does not move with them.

The difference between these two experiences is not intelligence. It is not virtue. It is loop length.

Agency lives in short loops.

That single sentence explains a great deal of human frustration.

In earlier centuries, many loops were short by necessity. You planted seeds, you saw the weather, you harvested. You built a fire, you stayed warm. You said a cruel word, you saw a face change in front of you. Feedback could be brutal, but it was immediate. Life was harsh, but it was legible in the sense that consequences arrived quickly.

Then human civilization grew in scale. Signals traveled farther. Decisions were distributed. Outcomes were delayed. That growth brought comfort and safety, but it also stretched feedback loops, and stretched loops weaken coupling.

When the loop stretches too far, a person can still be told what is coming, but they cannot steer.

So the first rule of preserving agency is to protect short loops wherever possible. In personal life that means choosing habits that give you fast honest feedback, like sleep, movement, writing, direct conversation, learning by doing. In a society it means building tools and institutions that reveal consequences early, that allow correction before branch points close, that keep reversibility available long enough to reach.

Reversibility is the second pillar.

In a finite volume, a reversible action keeps more futures reachable. It preserves branching. It preserves the sense of an open corridor. An irreversible action collapses branching. It locks a path. It turns prediction into narration.

Think of a simple digital example. You delete a file and it is gone instantly. That is an irreversible action. It creates panic, because the branch point closes in a moment. Now imagine the same action with a visible undo window. The file moves into a reversible state for a period of time. You can recover it. The panic drops, not because the mistake is impossible, because the future remained reachable for longer.

That is a tiny example, but it scales.

A payment above a threshold that waits in a reversible state protects people from scams without stripping freedom. A contract that includes a cooling period protects people from impulsive irreversible commitments without treating them as children. A critical medical decision that includes a second confirmation step and a clear undo path protects people from misunderstanding while keeping dignity intact.

The key is where the delay sits. Delay can steal agency or preserve it. A delay that blocks a lever until after the branch point closes is theft. A delay that creates space for reversal before the branch point closes is protection.

This is why reversibility is a civilizational invention. It is one of the few ways to preserve agency in large scale networks of delayed decisions. It is the way to keep the corridor from snapping shut instantly.

The third pillar is clarity.

A warning delivered early but written in vague language is late in practice. A lever offered without plain meaning is a lever that arrives after the mind catches up, and by then the window can already be closing.

In the Volumetric Time Model, meaning has a latency. Your mind does not absorb a sentence instantly. Your mind parses, interprets, compares, doubts, and only then acts. That is why clarity is not a luxury. Clarity is part of timing.

A simple sentence can widen the effective window because it reduces comprehension delay. A confusing sentence can shrink the effective window even if it arrives early. This is why the ethics of access includes plain meaning. The truth is not enough. The truth must arrive in a form that becomes usable inside the window.

This brings us to the fourth pillar, signal integrity.

A single urgent signal can be usable. A hundred polite reminders can destroy usability. A channel can be filled until nothing remains distinguishable. In that state, the cone of access narrows even as information volume increases. This is one of the central ironies of the modern world, more information can produce less access.

Agency requires not only signals, but signals with shape. Urgency must stand out. Options must be distinct.

Levers must not be disguised as decoration. If everything looks equally important, then nothing becomes actionable in time.

A fifth pillar is the reality of cost.

Agency is not free. Every widening of access requires energy, and not only electrical energy. It requires attention, time, and physical capacity. A person who is exhausted has a narrower cone. A person who is overloaded has a narrower cone. A person who is surrounded by noise has a narrower cone.

This means that preserving agency is partly an energy problem. It is partly a nutrition and sleep problem. It is partly a social support problem. It is partly a design problem. It is partly a fairness problem, because different people live with different attention budgets and different recovery budgets, and a narrow budget is easily exploited by long loops and late levers.

If you want a simple way to see this, imagine two lives with the same intelligence and the same values.

In the first life, the person has time to rest, time to think, time to recover, and access to clear signals. Their cone stays wide enough to overlap with branching. Their mistakes can be reversed. Their levers are reachable.

In the second life, the person is exhausted, overloaded, pressured by short deadlines and long delays, surrounded by noise, forced to decide under time scarcity. Their cone narrows. Branch points pass unnoticed. Reversibility expires before they can reach it. The person is later blamed for poor choices, but the real story is that the effective window was too small for a human being with a limited energy budget.

This is why a society that wants genuine agency must treat energy and recovery as part of access. If you drain people, you shrink their reachable futures. If you keep them permanently saturated, you make their decisions ornamental.

Now we can name the practical consequence of all these pillars. Agency is preserved by maintaining coupling through loop design.

You keep loops short.

You keep meaning clear.

You keep levers real.

You keep reversal available.

You keep signals distinct.

You keep the human energy budget in mind.

When these conditions are present, the future feels less obvious and more open. It can feel uncertain, but it is a fertile uncertainty. It is uncertainty that comes from branching that remains reachable. It is the uncertainty of a landscape with multiple roads, not the uncertainty of darkness.

When these conditions are absent, the future can feel obvious and closed. It can feel legible in a cold way. It can feel like narration. People often mistake that feeling for maturity. They say they are realistic. They say they are simply accepting. Sometimes they are. Often they are just living inside a corridor where the levers have been removed.

In the near future, the most important innovation may not be better prediction. Better prediction is easy to celebrate, because it looks like intelligence. The more important innovation may be better coupling. Better coupling keeps lives steerable. Better coupling keeps the corridor from closing too early. Better coupling keeps responsibility honest, because it ensures people are not blamed for failing to do what no human could do inside the available window.

In daily life, preserving agency will look surprisingly humble. It will look like fewer prompts, not more. It will look like clear instructions that arrive before the stakes lock in. It will look like undo windows for actions that change lives. It will look like direct human contact for decisions that cannot be reduced to a button. It will look like a culture that respects the difference between acknowledgement and consent.

Most of all, it will look like a new instinct in the mind.

When you are shown a forecast, you will not only ask whether it is accurate. You will ask whether you have a lever in time.

When you are offered a choice, you will not only ask whether the choice exists on the screen. You will ask whether the world will move with you if you choose it.

This is what it means to live inside a finite volume without surrendering. You do not deny the sphere. You do not demand impossible freedom. You protect the regions where coupling still exists, and you refuse to confuse late clarity with real power.

In the next part, we turn these pillars into a simple set of daily laws. Not commandments, not slogans, but practical rules for keeping your cone honest and your levers real, so that the life you traverse inside the volume remains steerable where steering is still possible.

A finished volume does not tell you what to value. It only tells you what is reachable. It tells you where the levers live. It tells you how quickly a door closes. It tells you how much of your life is shaped by timing. Once you see that, ethics stops being a debate about abstract virtues and becomes a question you can ask in the middle of an ordinary day. Does this choice couple. Does this warning arrive in time. Does this lever change anything. Does this path stay reversible long enough for a human being to use it.

In earlier chapters we talked about access like a cone, about branching like a forest of threads, about the paradox where prediction sharpens as options narrow. Now we turn that map into daily practice. Not motivational practice, not moral performance, not the theater of being a good person on a screen. Practical ethics, the kind that preserves agency where agency still exists, and that refuses to be seduced by late clarity.

These are laws in the humble sense. They are rules of thumb for living inside the volume. They are not perfect. They are not universal. They are honest.

Law one. Protect short loops.

Short loops are where life answers you quickly. Short loops let you correct course before the branch point closes. Short loops make error useful instead of fatal. You protect them in your body by sleeping, moving, eating well, and giving your mind silence, because exhaustion narrows the cone and delays meaning. You protect them in your relationships by speaking early, listening early, repairing early, because repair gets harder when resentment has had time to harden. You protect them in your work by choosing tasks that give you feedback you can actually use, not tasks that produce only delayed judgment. A short loop is not comfort. A short loop is power.

Law two. Treat late certainty as a warning.

When the future starts to feel obvious, do not celebrate. Ask what closed. Ask which reversals expired. Ask which levers were removed. Late certainty often arrives when the corridor has already narrowed. It can feel like wisdom. It can be surrender disguised as clarity. If you want a simple test, notice whether you can still change anything in the next hour, the next day, the next week. If you cannot, then certainty is not guidance, it is a report.

Law three. Do not confuse monitoring with steering.

Modern life loves dashboards. Modern life loves alerts. Modern life loves a constant stream of numbers that make you feel informed. But monitoring is not steering unless you have a lever in time. You can watch your finances collapse in high resolution and still be unable to intervene because the only lever is an appeal that runs too slowly. You can watch your health trend downward in beautiful graphs and still be unable to act because the appointment arrives after the window. You can watch a relationship drift and still be unable to repair because you waited until the words turned heavy. If you monitor, also ask what action you can take today that couples to the outcome. If the answer is none, then monitoring is not a tool, it is a mood.

Law four. Refuse ornamental choices.

A choice that cannot change the outcome is decoration. It is theater. It is a button that records compliance. You will be offered many ornamental choices in your life, because they are cheap to provide and useful to those who want you to feel responsible without giving you power. You learn to recognize them by their timing. The choice appears after the branch point. The choice appears with no clear lever. The choice appears as a preference that will be recorded, thanked, and ignored. You do not need to rage at these choices. You simply name them. You stop investing your attention in them. Attention is the steering wheel, do not give the steering wheel to decoration.

Law five. Spend energy where it buys coupling.

Energy is the currency of agency. If you are depleted, your cone narrows, your meaning latency grows, your errors multiply, and your levers become harder to reach. The world will happily take advantage of this. It will give you decisions when you are tired, it will give you fine print when you are rushed, it will give you important prompts when your attention is already saturated. So you treat your energy budget as a moral resource. You protect it. You stop spending it on noise. You stop spending it on performative outrage that changes nothing. You spend it on the places where action still couples, on the phone call that prevents a misunderstanding, on the walk that clears your mind enough to see the branch point, on the difficult conversation that repairs early, on the small planning step that keeps a future reversible.

Law six. Choose relationships with low latency.

This is one of the strangest conclusions of a time volume, love is partly a timing phenomenon. A relationship where feedback is fast and honest stays steerable. A relationship where feedback is delayed becomes a corridor. When you cannot get a clear response, you start predicting instead. You start building stories. You start living in late certainty. That is where decay begins. So you choose people who answer, not perfectly, not instantly, but within a human window. You choose people who say what they mean while it can still help. You choose people who can apologize before the apology becomes a receipt. You become that person as well.

Law seven. Build reversibility into your life.

You cannot make everything reversible. Some doors close. Some doors should close. But you can design your days so fewer doors close by accident. You keep savings as reversibility. You keep time margins as reversibility. You keep plans with options as reversibility. You keep friendships alive as reversibility. You keep your health cared for as reversibility. You keep simple backups. You keep a habit of leaving yourself a way out. This is not cowardice. This is respect for the geometry. In a finite volume, reversibility is how you keep branching reachable.

Law eight. Demand plain meaning for high stakes.

If the stakes are real, the language must be plain. If a prompt can change your money, your health, your home, your family, then you should not accept polite fog. Polite fog is a timing trick. It steals the effective window by slowing comprehension. You learn to ask for clarity, not as a complaint, as a requirement. You learn to rewrite complicated instructions into plain statements before you act. You learn to slow down when the words are slippery, because slippery words are often a sign that someone wants you to acknowledge rather than understand.

Law nine. Reduce noise to protect signal integrity.

Signal integrity is the condition that makes meaning usable. If you want your cone to remain honest, you must protect it from flood. You do this in simple ways, fewer prompts, fewer feeds, fewer constant updates, fewer channels that demand attention with no lever. You do not do this to become pure. You do it because noise is delay. Noise makes urgent signals look like everything else. Noise makes the brain numb. Noise trains you to tap without meaning. In a world that loves to drown you, silence becomes a form of freedom.

Law ten. Move upstream in time.

This law is the answer to the prediction paradox. When you notice late certainty, you do not fight the certainty. You move upstream. You act earlier next time. You set triggers that fire before the corridor forms. You decide while branches are still open. You practice repair while repair is still light. You place your own delays where they preserve reversibility. You build habits that keep your future less obvious and more open. This is why some of the wisest people look calm, not because they see everything, because they act before the corridor narrows.

Law eleven. Do not moralize what was not reachable.

Once you understand access, you stop turning every outcome into a story about character. You stop blaming yourself for not saving a future you could not reach. You stop blaming others for failing inside windows that were too small. This is not softness. This is precision. Responsibility lives where a lever was real and available in time. Guilt that ignores timing is not morality, it is superstition. The volume teaches you to be honest about what was reachable, so that you can focus your effort where effort still couples.

Law twelve. Keep one place where you are not optimized.

This is a strange one, but it matters. If every part of your life is tuned for efficiency, you will eventually shrink your own windows. You will turn yourself into a corridor. You will learn to treat every delay as unacceptable, and then you will accept delays that steal your agency because you are too tired to fight. So you keep one place that is slow, not slow because it steals, slow because it restores. A walk with no goal. A meal with no screen. A conversation with no agenda. A page you write only to hear your own mind. These are not luxuries. They are the places where your cone widens again, where meaning becomes clear again, where levers become visible again.

None of these laws make the sphere disappear. None of them give you control over the entire volume. That is not the deal of being human. The deal is smaller, and because it is smaller it is real. You can protect the regions where coupling still exists. You can refuse ornamental choices. You can demand reversibility. You can build short loops. You can spend energy where it buys steering. You can keep your future less obvious by keeping more futures reachable.

This is the ethical promise of the Volumetric Time Model. Not that you can rewrite the universe, but that you can stop being tricked by late clarity. You can stop calling receipts choices. You can stop calling forecasts power. You can learn to live inside the cone with skill, and to build a world where more people have access to real levers in time.

In the next part we face the boundary that no leverage crosses. The edge of traversal. The place where the path ends. Not as a moral verdict, as a physical condition.

Death As Boundary Condition.

Every path inside the volume ends. Not because the universe is cruel, but because finitude is part of what it means to be a physical object. A star has a lifetime. A mountain has a lifetime. A molecule has a lifetime. A thought has a lifetime. The Volumetric Time Model does not introduce death as an idea, it simply refuses to hide the boundary behind comforting metaphors. The boundary is there. Traversal ends there.

It is tempting to treat this as a philosophical statement, and many traditions have tried to turn the end of life into a moral verdict, a cosmic grade, a reward, a punishment, a story that makes the boundary feel earned. The VTM frame does something quieter and more demanding. It treats death as a boundary condition, a physical limit, a stopping point in a finite space of admissible histories. Not an insult. Not a surprise. Not a moral commentary from the sky.

If you want to feel this idea in your bones, look at the simplest physical example, the arc of a thrown ball. The ball leaves your hand, it rises, it slows, it turns, it falls, it stops.

The arc is finite. The end is not a failure. The end is the completion of traversal under gravity and air resistance and the constraints of a particular initial push. You can make the arc longer. You can make it shorter. You can add spin. You can change the angle. But you cannot throw the ball and demand it never land, because landing is built into the shape of motion in a world with friction and finite energy.

Life is not a ball, but the same honesty applies. You are a physical process moving through a space of possibilities with limited energy, limited time, limited access, limited reversibility. You do not traverse forever. The boundary is not optional.

This sounds obvious, yet it is rarely lived as obvious, because human minds are built to treat the present as permanent. We wake up and the world looks stable. The light returns. The air fills our lungs. The next hour feels guaranteed. The next week feels likely. The next decade feels normal. The boundary is distant enough that it can be treated as an abstraction, until it is not.

The Volumetric Time Model changes the emotional texture of this fact. It says, the boundary is part of the geometry, and the geometry matters because it shapes what access means. A finite volume is not only a constraint, it is the reason your access has meaning at all. If everything were infinite, urgency would dissolve into fog. If every path were traversable forever, branch points would not feel like branch points, they would feel like optional detours you could always revisit. In a finite volume, some doors close. Some windows end. Some reversals expire. That is why the timing of access matters.

This is one reason death matters to the end of this book, not as drama, but as calibration. If you ignore the boundary, it is easy to sell fantasies. It is easy to promise perfect prediction. It is easy to promise that more information will solve everything. It is easy to promise that you can optimize life into safety. It is easy to promise that every mistake can be fixed later.

The boundary makes these promises visible as what they are, promises that forget time is finite and that access is limited.

In physics, boundary conditions do not tell you the entire story, but they decide the shape of the story. A guitar string vibrates differently depending on where it is fixed. A wave in a drumhead behaves differently depending on the edge. A beam of light behaves differently in a cavity with reflective walls. The edges matter. They do not moralize. They constrain.

Death is the most personal boundary condition. It does not tell you which path you will take. It tells you that the path ends. The end shapes what can be reached, what can be reversed, what can be postponed, what must be done while a window is still open.

There is a second reason death belongs in this chapter. It is the place where the separation between prediction and steering becomes undeniable. There are limits no lever crosses. There are processes no warning can undo once they pass a certain point. There are trajectories where the reachable set shrinks no matter how much clarity you gain. This is not pessimism. It is the truth of finite physical beings.

If you have ever watched a distant star through a telescope, you have already lived inside the VTM logic without naming it. That star may have died long ago. The light reaching you tonight is an old letter, delayed across space. You are receiving information from a past that still exists as a record in the present, carried by photons. You can learn from it. You can measure it. You can predict what kind of star it was. You cannot steer it.

This is a cosmic version of forecasting without power. It is not tragic. It is simply the separation of access from influence at large distances and long delays.

Now bring that cosmic lesson back to the human scale. A diagnosis can arrive as information, and still the levers may be limited. A warning can arrive as a trend, and still the branch points may already be narrowing. A person can see their own patterns with increasing clarity, and still be unable to correct them if the loop is long and the reversibility is gone.

The boundary does not mean you are helpless. It means that leverage is local. It means that the meaningful work of living happens in the regions where the world can still answer in time. It means you should not confuse late clarity with early power.

This is where the VTM frame becomes unexpectedly humane. Many people carry regret as if regret is proof of moral failure. They replay their life and say it was obvious, I should have known, I saw it coming. They treat the clarity of hindsight as evidence that a different path was reachable at the time. Often that is false. The path exists in the full volume, but it was not reachable from the position they inhabited with the access they had.

The boundary condition helps with this, because it forces you to separate two questions that humans often blend into one. What could have existed in theory. What could have been reached in practice.

The first question is infinite and cruel. The second question is finite and honest.

To live inside a finished volume without becoming numb, you need this honesty. You need to treat your life as a traversal of reachable paths, not as a courtroom where every outcome is judged by a fantasy of perfect access.

This does not remove responsibility. It relocates it. It says, responsibility lives where a lever was real and reachable in time. Responsibility does not live in corridors where the levers never existed. Responsibility does not live in windows that were too narrow for a human body and a human mind.

Death also reshapes a quieter part of ethics, the ethics of attention. If time is finite, then attention is finite. Every minute you spend on ornamental control is a minute you cannot spend on a real lever. Every hour you spend inside noise is an hour you cannot spend protecting a short loop. Every day you spend waiting for a late warning is a day you could have spent building earlier coupling.

The boundary does not demand panic. It demands selection.

This is why the last chapters of this book keep returning to small practical rules, protect short loops, demand reversibility, keep signals clear, move upstream in time. These rules are not self improvement slogans. They are ways of keeping more futures reachable before the boundary arrives.

There is another cosmic point worth stating plainly. The universe does not need your story. Stars do not care about your plans. Galaxies do not adjust their rotation to your hopes. Yet it is also true that the universe produces regions where local levers exist, where small actions couple to outcomes, where meaning can be created because a loop can close. Life is one of those regions.

The VTM frame does not remove meaning. It explains what meaning depends on. Meaning depends on coupling. Meaning depends on the fact that you can act and the world can answer. Meaning depends on the fact that the answer arrives soon enough to matter. Meaning depends on the fact that not all paths remain reachable, so choosing is real.

In that sense, finitude is not the enemy of meaning. Finitude is the condition that makes meaning possible.

A finished volume can be terrifying if you imagine it as a prison. A finished volume can also be liberating if you understand what it actually says. It says, you do not have to pretend you control everything. You do not have to pretend you can see everything. You do not have to pretend you can optimize away every risk. You can stop buying the sedative promise of perfect prediction. You can stop treating information as if it were leverage.

You can focus on the honest levers you do have.

This is how the chapter should land emotionally. Not with denial of death. Not with melodrama about death.

With a calm recognition that death is the boundary of traversal, and that this boundary makes the timing of access matter. It makes reversibility valuable. It makes signal integrity valuable. It makes short loops precious.

And it leads to a final thought that prepares the last part of the book. If the boundary is real, then the ethical project is not to demand infinite access. The ethical project is to protect access where it can exist, to make it honest, to keep windows open early, to keep levers real, to keep reversibility alive, to keep more futures reachable for more people before the corridor narrows.

That is the kind of optimism that fits physics. It is not the optimism of a guarantee. It is the optimism of craftsmanship.

The Final Page

Return to the sphere.

Not the museum sphere now, not the teaching prop. The real one that exists as an idea inside the reader, the image that should remain after the last sentence fades. A finite volume of admissible histories, dense with threads, quiet with possibility, bounded at the edge where traversal ends.

Inside that sphere there is a moving point. That point is you.

It is not heroic. It is not special in the cosmic sense. It does not glow with destiny. It simply moves, and as it moves it lights a narrow cone ahead, a region of access shaped by distance and noise and attention and delay, shaped by the timing of signals and the placement of levers, shaped by the human fact that meaning has latency and that action has limits.

That cone is your life as you actually live it.

Most of the sphere is never seen. Most of the threads are never touched. Not because they are not real in the full volume, but because reach is always local. Your world is not the entire space of admissible histories. Your world is the overlap between access and branching, the region where a lever can still couple to a consequence before the branch point closes.

This is the ending the Volumetric Time Model earns. Not a grand metaphysical verdict, not a surrender to fate, not a promise that you can rewrite the universe.

A sober claim.

Globally consistent reality can coexist with locally limited access.

And a second claim that feels almost harsher because it is so practical.

Prediction can persist after influence collapses.

This is what we have called forecasting without power, and by the end of this book it should no longer feel like a technical phrase. It should feel like a recognizable pattern in human life. The future can become legible at the exact moment it becomes unreachable. Late clarity can be a symptom of narrowing, not a gift. A clean forecast can be offered as a substitute for a real lever. A button can be offered as a substitute for consent.

Once you see that pattern, you cannot unsee it. And that is precisely why the ending cannot be despair.

Despair is what happens when you confuse a view with a grip.

The VTM frame separates them. It says, a view is not a lever. A forecast is not an action. A warning is not protection unless it arrives early enough and in plain meaning and with a real lever inside the effective window. A choice is not a choice unless the world will move with you when you choose it.

This is not poetry. It is timing.

You can measure it. You can audit it. You can design around it. You can build a civilization that is honest about it.

That is why the most important moral idea that emerges from a volumetric universe is not about freedom in the abstract. It is about access in time.

Responsibility tracks operational accessibility.

Not what was true in the full sphere.

What was reachable through the cone, with the windows that actually existed, with the levers that were actually available.

This simple shift removes a great deal of superstition from human judgment. It removes the fantasy that people are gods who should have known, should have acted, should have been perfect inside seconds. It relocates accountability upstream toward those who control timing and gate levers and flood channels with noise. It makes fairness a matter of window design rather than moral theater.

You can now see why the book had to end with this chapter. Everything we have discussed leads here.

A universe can be fixed without being meaningless.

A life can be finite without being trivial.

Agency can be real without being infinite.

Agency is not the ability to rewrite the sphere. Agency is the ability to steer inside the regions where coupling remains above threshold. It is preserved by short feedback loops, by clear meaning, by real levers, by reversibility, by signal integrity, by respect for the energy budget of a human mind.

In other words, agency is preserved by craftsmanship.

This is the kind of hope that fits physics. Not the hope of guarantees. The hope of design.

In the coming decades, the world will keep getting better at prediction. That is the easy part. Sensors will improve. Models will refine. Forecasts will sharpen. Risk curves will become more precise. The future will become more legible in many domains.

The danger is that this legibility will be confused with empowerment. The danger is that people will be shown the future in high resolution while the levers are quietly removed, delayed, gated, buried, or made ornamental. The danger is a world where everyone is informed and few can steer.

If that happens, the emotional landscape of society will change in a recognizable way. People will feel haunted by foreknowledge. They will feel guilty for outcomes they could not reach. They will be asked to acknowledge, to confirm, to accept, to comply. They will be blamed as individuals inside corridors designed to be irreversible.

The Volumetric Time Model offers a different future.

A future where we treat access as safety critical. A future where we do not confuse late certainty with fairness. A future where we demand effective windows, real levers, and reversibility. A future where timing receipts make invisible delay visible. A future where standards for access can be audited and captured less easily because tests are adversarial and public.

This is not utopia. This is mitigation. This is the difference between a world that quietly steals reach and a world that keeps more futures reachable for more people.

And now the ending must become intimate again, because the reader closes the book alone.

So here is the final practice, the simple test you carry with you, the one that turns the Volumetric Time Model from an idea into a daily instrument.

When you are shown a forecast, ask what lever you have.

When you are offered a choice, ask whether it couples.

When you are handed a button, ask whether it moves the world.

When you feel late certainty, move upstream in time.

Protect short loops.

Protect reversibility.

Protect signal integrity.

Do not spend your attention on ornamental control.

Do not moralize what was not reachable.

If the universe is a volume, your life is a traversal. Your dignity lives in the honesty of that traversal. Not in fantasies of total control. Not in the sedative comfort of perfect prediction. In the real work of keeping your cone honest and your levers real.

Return to the sphere one last time.

The moving point continues. It does not need to know the whole sphere to make meaning. It only needs enough access to steer where steering is possible. It only needs enough reversibility to keep branching alive. It only needs enough clarity to act before the branch point closes.

The boundary remains. Traversal ends. That is not a curse. That is the edge that makes choices real. And in the final image, the one that should remain after the last page, the cone of light is narrow but honest, sweeping forward as the point moves.

Not infinite.

Not omniscient.

Not powerless.

Just human, inside a finished universe, still able to reach what is reachable, still able to build a world where more of life remains within reach.

Existence does not guarantee access.

But access can be protected.

And that is enough.

FORMULA DECK (Chapter 11)

Living in a Finished Universe, Levers, Windows, and the Boundary

1. **Traversal as a worldline through admissible histories.**
   Let (M, g) denote spacetime with metric g. Fix a compact region $V \subset M$ with boundary ∂V, and

let H be the admissible history set on V. A particular realized world corresponds to a history ϕ in H. An embedded observer corresponds to a worldline γ through V. The local state experienced by the observer at time t is a state along ϕ at the location $\gamma(t)$.

2. **Boundary condition and termination as a stopping time.**
   Define a stopping time T for the observer trajectory, interpreted as the end of traversal. Termination occurs when the worldline reaches the boundary ∂V, or more generally when it reaches a designated terminal subset $B \subset V$. This encodes death as a boundary condition on traversal, not as a moral or evaluative claim.
3. **Access as a delayed noisy channel.**
   The observer does not access the local state directly. Instead the observer receives an observation stream generated through a delayed and noisy channel. Observations reflect past states with latency τ and are corrupted by noise. As a result, local access is constrained even when the global history ϕ is internally consistent.
4. **Prediction coupling as readability without influence.**
   Prediction coupling measures how readable the future is from present observations. Large prediction coupling at delay τ indicates that future states are statistically predictable from current observations. This readability does not

imply that the observer can influence those future states.

5. **Actions and steering coupling.**
Let U denote an action applied by the observer. Local evolution depends on the current state, the applied action, and uncontrolled disturbances. Steering coupling measures whether actions measurably influence future outcomes beyond what is implied by observation alone. If steering coupling is indistinguishable from a null baseline, actions do not provide real leverage.
6. **Agency horizon as an operational threshold.**
Fix a practical near-zero threshold for steering, or a null-calibrated threshold obtained from shuffle or shift controls. The post-zero agency horizon is the smallest delay at which steering coupling falls below this threshold. This horizon marks the operational limit of influence, not a metaphysical boundary.
7. **Forecasting Without Power region.**
The FAWP regime consists of delays where prediction coupling remains significant while steering coupling is below threshold. In this regime the future is readable but not reachable. This formalizes the experience of seeing what is coming without having the ability to change it.
8. **Effective window for a lever.**
Let the signal time be the moment a consequential signal becomes available. Let the lever time be the moment a relevant lever

becomes available. Let the branch close time be the last moment when the outcome can still be altered. The effective window length is defined as the branch close time minus the later of the signal time and the lever time. A usable lever requires this window to be positive. If it is zero or negative, the lever is operationally nonexistent for that event.

9. **Reversibility window and undo horizon.** Some outcomes can be reversed for a limited time. Let the undo horizon be the end of this reversal period. The reversibility window length is defined as the undo horizon minus the later of the signal time and the lever time. A domain-specific minimum reversibility requirement encodes protection through reversibility rather than removal of options.
10. **Timing receipt as a minimal auditable record.** Represent a consequential interaction by a timing receipt consisting of the signal time, lever time, branch close time, and undo horizon. From this receipt one can compute effective and reversibility windows without storing content. This enables accountability about timing and leverage while minimizing exposure of personal details.

Prediction and Steering Coupling Limits

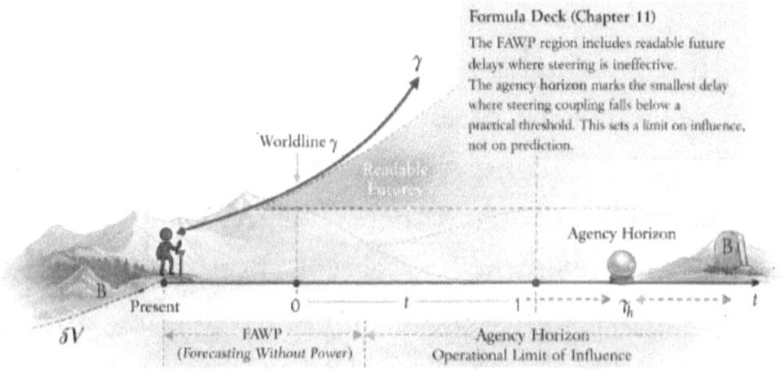

FAWP (*Forecasting Without Power*)
Readable but not Reachable

Agency Horizon
Operational Limit of Influence

FAWP (*Forecasting Without Power*) – Readable but not Reachable

Agency Horizon = Operational Limit of Influence

Epilogue

On a clear night, you can walk outside and look up, and it feels like the oldest kind of certainty. The sky is full of points of light, steady and calm, as if the universe is showing you what is there. But the calm is a trick your senses have learned to accept. Every star you see is a message that arrived late. Some arrived minutes late. Some arrived years late. Some arrived so late that the source may have changed shape, or faded, or vanished, long before your eyes received the news.

This is not a poetic flaw in astronomy. It is a basic fact about access. Light takes time. Signals take time. Information has a speed. And because you live inside that speed limit, the world you experience is never the world as it is in some global sense. It is the world as it can be reached.

You already know this in daily life, even if you never say it out loud. You have looked at a message and felt the delay in your chest. You have watched a situation turn while you were still reacting to an earlier version of it. You have seen a trend, in your health, in a relationship, in a plan, in the mood of a crowd, and you have known what was coming, and still missed the narrow window where a small move could have changed the outcome. Later you might say, I saw it coming, and you might sound like you are trying to protect your pride. But the feeling is real, and it has a clean structure. It is the feeling of being correct without being able to steer.

We are taught to treat knowledge as power. It is a comforting slogan. It fits on posters. It sounds like progress. But it is only partially true, and the missing part matters more than we like to admit. Knowledge is power only when the channel between knowing and doing remains open, fast, and clean. When that channel is delayed, noisy, or narrow, knowledge becomes something else. It becomes prediction without leverage. It becomes the ability to watch a wave arrive from far away, and still be unable to move the boat.

The modern world gives you a strange mix of abundance and helplessness. You have dashboards, notifications, summaries, forecasts, and warnings. You have more measurement than any generation in history. And yet you may feel less able to move certain outcomes, not because you are weaker, but because the timing of influence has changed. You can learn early, but you cannot act early. You can see clearly, but you cannot reach. You can predict, but you cannot steer.

This book is built around a simple split that sounds obvious once you say it, but changes everything once you take it seriously. Existence is not access. A fact can be real and still be unreachable to you. A pattern can be present and still be invisible to your tools. A future can be fixed in the broad geometry of reality and still remain unavailable to your decision making in time to matter. None of this requires mysticism. It requires only that you are a local observer, inside a world where information travels and gets filtered, compressed, delayed, and sometimes lost.

Imagine a library that contains every version of a story, every possible route the story could take, every chapter that could be written. The library exists. But you are not holding it in your hands. You are allowed to read only one thin page at a time, and the page arrives with smudges, missing lines, and a time lag. You can infer the plot. You can guess the next chapter. You can sometimes guess very well. But the existence of the library does not mean you have access to the shelf you want. Now imagine that the story is your life. Imagine that time is not a line you ride like a train. Imagine that time is a volume, a bounded region with structure, and your lived experience is a path through it. The volume contains many admissible histories, many ways the world could be consistent with its own laws and constraints. You do not create the volume by choosing. You move through it. Your choices matter, but not as magic that summons new realities from nothing. Your choices matter as steering, selecting direction within what is already consistent.

This picture, time as a volume, is the Volumetric Time Model. It is not a metaphor. It is a way to organize what physics already hints at, and what daily experience already proves. The universe can be globally consistent, and still feel uncertain locally. A full shape can exist, while you only receive a thin slice. This is how you can have determinism without omniscience. This is how you can have reality without full reach. This is how you can be surrounded by true facts that do not help you. There is a second split that follows immediately, and it is the heart of the argument. Forecasting and steering are not the same. Forecasting is about building an internal picture that tracks the world. Steering is about pushing the world onto a different path. Forecasting can survive delays that would destroy steering. In fact, forecasting sometimes improves in the very conditions that make steering collapse, because delays often come with smoother signals and bigger patterns. You can be excellent at predicting a storm while being unable to change the day it hits. You can be excellent at predicting a market swing while being unable to execute in time. You can be excellent at predicting your own exhaustion while still failing to prevent it, because the prevention required a move long before the exhaustion became obvious.

Once you see this, the phrase I saw it coming takes on a new meaning. It becomes a report about channels, not a confession about character. The forecasting channel was open. The steering channel was closed, or too delayed, or too noisy to matter.

The model in this book does not ask you to surrender to fate. It asks you to become precise about where leverage lives. Leverage is not a feeling. It is not confidence. It is not moral worth. Leverage is a measurable coupling between action and outcome under the constraints of time, noise, and bandwidth. And like any measurable thing, it has boundaries. There is a region where small actions can steer. There is a region where actions still exist but their effects drown in delay and randomness. There is a boundary where the steering channel collapses even as prediction remains. That boundary is what we will call the Agency Horizon.

If you hear the word horizon and think it sounds philosophical, good. It should. But it is not only philosophy. In the chapters ahead, we will treat this horizon as a scientific object. We will build it from simple ideas, then test it with experiments, then show how it explains a puzzling regime that looks like a contradiction until you name it clearly.

That regime is Forecasting Without Power. It is the situation where prediction stays strong beyond the point where influence is indistinguishable from noise. It is the experience of being right while being unable to steer. It is the experience of living in a world that keeps sending you accurate postcards from the future of your own trajectory, while denying you the timing needed to change the address. If that sounds bleak, it does not have to be. Clarity is not despair. Clarity is the beginning of sane design, sane ethics, and sane self understanding. When you confuse prediction with power, you blame people for outcomes they could not reach in time. You punish after the fact and call it responsibility. You praise early warning and pretend it implies control. You build tools that watch, and assume watching is doing. You congratulate yourself on foresight, and miss the true question, which is where the steering channel actually lives.

The Volumetric Time Model gives you a way to keep two truths in the same mind without forcing them to fight. Reality can be globally consistent, and your access can still be limited. Prediction can persist, and influence can still collapse. Your choices can matter, and your reach can still have strict boundaries. Once you stop treating these as contradictions, you can start measuring them, and once you can measure them, you can stop arguing in circles and start designing around what is real.

This prologue is an invitation to look at time with a new kind of humility, and a new kind of ambition. Humility, because you will see how much of your life is lived on delayed records and partial views. Ambition, because you will also see that leverage, when it exists, is not mystical, it is physical, and it can be protected, extended, and audited. We will build the model carefully, chapter by chapter, with everyday examples, with cosmic intuition, and with an insistence on what can be tested. The formulas will be there when we need them, at the end of each chapter, as a compact deck for readers who want the underlying structure. But the main story will remain simple. Time is not just something you pass through. It is the medium of access. It shapes what you can know, and when you can act, and whether knowing can become doing. When you feel powerless, it is often because your steering channel has been pushed past its horizon. When you feel certain, it is often because your forecasting channel still has signal. This book is about learning to tell the difference.

Author Commentary

There is a quiet feeling many people recognize now, even if they have never named it.

You find out what is coming, and you still cannot change it. The warning arrives. The forecast is accurate. The trend line is clean. The numbers are persuasive. And yet the moment you needed has already passed. The outcome proceeds anyway, almost politely, as if the universe is saying, yes, you are correct, and no, it does not matter.

This book exists because that feeling has become common.

It used to be that uncertainty was the main problem. A storm might arrive, but no one knew where. A patient might decline, but no one saw it early. A supply chain might snap, but no one noticed the weakness. In those worlds, the tragedy was ignorance.

In the modern world, the tragedy has shifted. Now the tragedy is often late clarity.

We are surrounded by tools that can see forward. Weather models, medical risk scores, market indicators, infrastructure monitoring, recommendation engines, even our own pattern recognition after years of experience. In many corners of life, the future is increasingly readable. The puzzle pieces are visible earlier than they used to be.

So why do so many people feel less free.

A popular answer is psychological. People are distracted. People are weaker. People are overwhelmed. People hesitate. People need discipline. There is a grain of truth in that answer, but it misses the deeper structure. It treats the symptom as the cause.

The deeper structure is timing.

Imagine spacetime as a completed landscape, a four dimensional atlas rather than a story written line by line. This is not mysticism, it is the normal implication of relativity. There is no single universal "now" floating through the cosmos like a spotlight. Instead, each observer carries a narrow moving beam of access along their own worldline. The universe does not have to "wait" for anyone's awareness to be real. The events are there in the atlas. What changes for an embedded observer is not the existence of events, but when and how information about them arrives.

Now put that physical picture next to ordinary life.

Most of what feels like freedom is the experience of short loops. You turn a wheel and the car responds. You touch a hot pan and pull away. You say something unkind and see a face change. You make a small correction and the world answers fast enough that you can correct again. Short loops are where agency lives.

But modern systems, in the everyday sense of the word, often live in long loops. Signals arrive late. Decisions take time. Actions land after conditions have changed. Reversals are difficult. Meaning is buried under noise. When loops get long, two strange things can happen at once.

The future can become more predictable, while becoming less steerable.

This is the central idea of the book, in plain language.

Prediction improves when the world becomes more instrumented and more constrained. When there are more sensors, more data, more models, more history, more pattern learning, the future becomes easier to infer. Also, when options quietly disappear, the future becomes easier to infer for a different reason, there are fewer exits left. Corridors form. Endings become legible.

Steering requires something else. It requires timely handles. It requires that an action taken now still couples to the outcome later, before the door closes. If that coupling dies, the human being can still act and still choose, but the world no longer answers in time.

In that situation, more forecasting does not necessarily help. Sometimes it harms. It extends the duration of helpless awareness. It turns knowledge into stress. Stress burns attention. Burned attention delays recognition. Delayed recognition shrinks the already thin window of real choice. The loop becomes even longer from the inside.

This is where Forecasting Without Power appears.

It is the condition where the future stays readable after influence collapses.

The book treats this as something that can be tested, not merely felt. That is why the experimental chapters matter. They are an attempt to build a clean map of a boundary, the point where steering drops to a null floor even while prediction remains substantial. Once that boundary is visible, the conversation changes. It becomes less about blaming individuals for not acting, and more about protecting the timing conditions that make action meaningful.

That shift matters because modern life is full of moral arguments that ignore time.

A person is shown a warning and later criticized for "not doing anything." An organization displays a forecast and later claims it was transparent. A platform notifies users and later says they were informed. Again and again, information delivery is treated as if it automatically creates responsibility.

But responsibility without reachable access is not responsibility. It is superstition wearing modern language.

If a lever is not reachable inside the effective window, it is not a real lever for that moment. If a signal is buried so deeply that recognition arrives late, then in operational terms the signal arrived late. If undo is impossible, early action becomes too risky, so people wait, and waiting becomes another delay tax.

That is why a theme that might sound technical, windows and reversibility, becomes deeply human. A wide undo window is not a luxury. It is a protection for real people, because it makes early action safe. A clean signal is not an aesthetic preference. It is a protection for real people, because it makes recognition fast. A short loop is not just efficiency. It is a protection for real people, because it keeps the world answerable.

This book is not asking readers to accept a finished universe as a gloomy verdict. The atlas picture is not meant to remove dignity. It is meant to locate dignity in the right place.

Dignity is not the fantasy that everything is controllable. Dignity is the craftsmanship of protecting the small regions where control still exists.

That craftsmanship is often boring. It looks like moving levers upstream. It looks like removing friction from urgent actions. It looks like making defaults reversible. It looks like reducing noise so meaning arrives on time. It looks like respecting human energy so attention does not collapse. But boring is not a flaw. Boring is often the shape of real freedom.

So the hope of this book is not that the future will become less knowable. The future will become more knowable. That is the direction of science, of sensors, of models, of measurement.

The hope is that we do not confuse knowable with steerable.

The hope is that we stop using forecasts as anesthesia.

The hope is that we learn to ask, every time a warning appears, not just what is likely to happen, but what can still be changed, and whether the handles are real, and whether the window is still open. If there is a single sentence that the deluxe edition commentary wants to leave behind, it is this. A forecast is not a handle. A handle is a handle.

And the future only becomes humane where handles arrive in time.

Coda 1

The Receipt

Before you close this book, there is one thing left to do. It will take ten minutes. It will be uncomfortable in a way that is hard to predict in advance, because the discomfort does not come from the exercise itself. It comes from what the exercise finds.

Pick one area of your life where you already know what is coming.

Not a vague anxiety. Not a general sense that things could be better. Something specific. A relationship that has been cooling in a way you have been careful not to name out loud. A number on a lab report that has been moving in a direction you have been meaning to address. A job that has been wrong for longer than you have admitted, in the quiet way that bad fits are wrong — not dramatically, not with a clear event to point to, just a steady, accumulating mismatch between who you are and where you spend your hours. Pick the one you already know about. The one that does not require research. Now write down the date you first noticed.

Not the date you were sure. Not the date you told someone. Not the date it became undeniable or dramatic or officially a problem. The date you first noticed. The first time a small signal crossed your threshold. The first time you felt the thing that you then set aside, smoothed over, filed under not yet.

If you cannot name a date, name a season. A year. An approximate location on your timeline. The precision is not the point. The honesty is the point.

Now ask the second question, and this one is harder.

What was the earliest real lever?

Not the lever you wish you had. Not the intervention that would have required a different life, a different income, a different relationship structure, a different version of yourself with more courage or more information or more time. The lever that actually existed. The conversation that was possible. The appointment you could have made. The boundary that was available to you. The small adjustment that the situation was still accepting before it hardened into something that required much larger force to move. Write that down too.

Did you use it?

If you did, write down when. If you did not, write down why not. Not a verdict. Not a confession. Just the record. What the timing actually looked like, laid flat on the page.

This is the receipt.

It is not meant to hurt you. It is not meant to produce guilt, which is one of the least useful emotional states available to a person trying to understand a system. Guilt looks backward with heat but without light. The receipt looks backward with light and no heat. It shows you the geometry of a situation you lived through, so that the geometry becomes visible rather than felt.

And here is what the receipt usually shows, once you stop looking away from it.

The signal arrived earlier than the action. Almost always. The gap is rarely dramatic, rarely the kind of thing that makes a clean story about negligence or failure. It is usually modest. A few weeks. A few months. Sometimes longer, but often not. The signal was there. It was even noticed. And then the gap happened — the gap between noticing and acting — and in that gap, something closed. A branch point passed. The corridor narrowed.

The lever that was available in March required more force by June, and by September it was no longer the same lever at all. The situation had moved forward, the way situations do, indifferently, without waiting for you to feel ready.

This is not a story about weakness. It is a story about how access works in time. You were not negligent. You were embedded. You were a local observer inside a world that produces signals before it produces obvious crises, and you were managing the hundred other signals that were competing for your attention at the same moment, and the signal in question did not arrive with a red banner and an alarm.

It arrived quietly, the way early signals do, which is exactly why they arrive early.

The system cannot give you a dramatic warning and an early warning at the same time. The dramatic warning is late by definition. The early warning is subtle by definition. You cannot have both. What you can do — what the receipt makes possible — is close the gap the next time.

Not eliminate it. The gap is a physical property of being a local observer with finite attention and finite energy inside a world that moves forward without asking. The gap cannot be reduced to zero. But it can be shortened. And shortening it, even by a little, changes what levers are available to you. This is the practical output of everything this book has been building.

Not a theory to carry around. Not a vocabulary to deploy at dinner parties. A discipline. A habit of taking receipts. A willingness to look at the timing of your own life with the same honesty that a good instrument looks at a signal — not to judge, not to punish, but to calibrate.

To know where your leverage actually lives.

To protect the loops that still answer.

To act inside the windows that still exist, before they become the windows that existed.

The book began in a hospital, at 2:13 in the morning, with a warning that arrived early and a lever that arrived late. That gap was not personal. It was structural. It was the shape of a system under constraint. Your life is also a system under constraint. The constraints are different. The channels are different. The signals are quieter and the stakes are smaller and the timelines are longer. But the structure is the same. The future that is already becoming visible does not have to be the future that you watch arrive. But it will be, unless the lever is real and the window is open and you are willing to use both before the branch point closes.

You already know which area of your life this is about.

You knew before you picked up this book.

Write down the date you first noticed.

Start there.

Coda 2

The Week of Writing It Down

Your body is a better instrument than you give it credit for.

Not a perfect instrument. Not an oracle. Not a thing to be trusted unconditionally or followed without question. But an instrument — a physical device that samples the world continuously, processes signals below the level of conscious attention, and produces outputs that are worth reading.

The output you have probably been ignoring is the one that arrives as vague dread.

Not panic. Not the sharp alarm of an obvious threat. Not the kind of fear that comes with a clear object and a clear cause. The other kind. The quiet kind. The unease that surfaces at odd moments — in the middle of an ordinary morning, in the pause between tasks, in the few seconds before sleep arrives — and that you have learned, through practice and necessity, to suppress before it becomes inconvenient.

You suppress it because it has no name yet. Because it is not actionable in its current form. Because the day is full and the feeling is vague and there are other things to attend to, and the feeling will probably turn out to be nothing, and you have been wrong about these feelings before, and you cannot spend your attention on every whisper.

This is a reasonable pipeline. It is also one that silences early signals at exactly the moment they are most valuable.

Here is the exercise.

For one week, every time you feel vague dread or unease — every time that nameless low-frequency signal crosses your threshold — write it down. Not an essay. Not an analysis. One sentence, or three words, or a timestamp and a location and a fragment. Whatever is true in the moment. Write it down before you talk yourself out of it or file it under later or wait to see if it grows into something more defensible.

Do not interpret it yet. Do not rank it. Do not decide in the moment whether it is a real signal or background noise. That judgment is exactly what the exercise suspends. Your pipeline has been making that judgment automatically, in real time, and some of what it has been classifying as noise is not noise. You will not know which is which until you have a list.

At the end of the week, look at the list.

You will find several categories, and they will be instructive in different ways.

Some entries will look small and strange in retrospect. The unease that turned out to be hunger. The dread that dissolved when you slept. The worry that was genuinely nothing — a misfire, a false alarm, a pipeline artifact. These are not embarrassments. They are calibration data. They tell you what your baseline noise floor looks like, which is information you need before you can identify anything above it.

Some entries will look like moods. Generalized fatigue wearing the costume of specific dread. A bad afternoon that your pipeline attached to a convenient object. These are also worth knowing. When your instrument is degraded by exhaustion or stress or overload, it produces more false positives. The list will show you when that happened, and you may notice a pattern.

And then there will be the other entries.

The ones that look different when you read them back. The ones that named something real, early, in a form you were not ready to accept at the time. The relationship. The situation at work. The thing you have been telling yourself is fine. The signal that arrived weeks ago in the form of a feeling you could not defend with evidence, and that has since grown into something more obvious and less movable. Those entries are the point of the exercise.

Not because they prove your intuition is always right. It is not. Not because they suggest you should have acted on every whisper. You should not. But because they show you something concrete about your own pipeline — about where your instrument is actually sensitive, about how far in advance the early signals arrive before they become undeniable, about the gap between first detection and acknowledged reality in your own life.

That gap is your personal agency horizon.

Not the theoretical one from the formal model. Your actual one. The one measured in your own weeks and months, in the specific domains where you live. The distance between when your body knows and when your mind admits. Between when the signal crosses the threshold and when you let it become actionable.

Most people, on seeing their list at the end of the week, discover that the gap is longer than they thought.

Not because they are bad at processing information. Because they are good at suppression. Because suppression is a skill that modern life rewards. The world does not make space for undefended unease. There is no meeting agenda item called early signal of unknown origin. There is no social format for saying I feel something is wrong but I cannot tell you what yet. So the signal gets smoothed. The unease gets filed. The pipeline continues producing clean output from noisy input, because clean output is what the environment asks for.

The list asks for something different.

The list asks you to stop preprocessing. To let the raw output sit on the page for a week before your interpretation layer gets to it. To treat your own discomfort the way a good instrument treats a suspicious reading — not by dismissing it, not by acting on it immediately, but by recording it carefully and returning to it later with more context.

This is the pipeline audit in daily life.

It does not require a laboratory. It does not require a model, or a formal framework, or a vocabulary you did not have before this book. It requires only a week and something to write on and a willingness to notice what you have been not-noticing.

At the end of the week you will know two things you did not know before.

You will know what your noise floor looks like — how much background unease your system generates that corresponds to nothing real. And you will know which signals you have been suppressing that deserved to be heard earlier.

The second kind is where your levers are.

The unease you feel before the situation becomes obvious is the world offering you access before the branch point closes. It arrives early because the signal is weak. It is weak because it is early. That is not a flaw in the instrument. That is the instrument working correctly.

The only question is whether you are listening.

Start the list tonight.

Not tomorrow. Not after you finish whatever is next. Now, before the book closes and the feeling you have been carrying quietly underneath the reading of these pages gets filed away under later, under not yet, under probably nothing.

Write it down. It is probably nothing.

It might not be.

That is exactly the difference the week is designed to measure.

Coda 3

Navigating VTM

Imagine your life as a shape, not a stream. Not a ticking clock, but a finished object you move through. From far enough away it is all there at once, like a coastline you could trace with a finger. From inside it arrives as weather, gusts of certainty, long humid stretches of not knowing, a sudden cold front of news that was already true before it reached you.

That is the split this book has been naming. A world can be globally consistent and still feel personal and unfair, because you do not live in the whole thing. You live in what reaches you. You live in what arrives in time to be used. You live inside a pipeline, and the pipeline has delay. It has noise. It has missing pieces. It has filters that smooth sharp edges into something your nervous system can survive. It has bottlenecks that do not care how urgently you want the answer. It has costs you do not see until after the fact, when you realize the price was paid in time.

This is why the future can look obvious and still be unreachable. The signal can be real. The warning can be real. The pattern can be real. And the lever can still be out of range.

If you have ever felt that, you already understand the emotional core of the Volumetric Time Model, even if you never use the phrase again. You have felt the strange humiliation of being correct without being able to act. You have felt the ache of recognizing a trajectory and then watching yourself move along it anyway, not because you are weak, but because the window for changing it was smaller than the story your mind tells about choice. You have felt what it is like to be shown a door after it has closed, and then be judged for not walking through it.

Everyday folks know this in their bones. You do not need a seminar to learn it. You learn it in relationships, when you see a slow drift becoming a breakup, and you cannot find the moment where a clean sentence would have changed the outcome. You learn it in health, when the body whispers for months, and then one day it stops whispering and starts shouting, and suddenly there is a calendar of appointments and a stack of instructions that arrive too late to be comforting. You learn it in work, when the mood turns, when a budget tightens, when a door quietly narrows, and you can feel the direction of travel while everyone else pretends it is not happening. You learn it in money, when the warning signs were present, and the action was possible, but the action required a kind of calm you did not have at the moment you needed it.

So let the coda be gentle about that. Let it remove the shame. There is a cruelty in hindsight that is dressed up as wisdom. It shows up as a smug sentence, you should have known. It shows up as a confident tone, the signs were all there. It shows up as a story where the only missing ingredient was your courage. Sometimes that story is true, and sometimes it is a lie that protects the world from admitting how much power is really timing and access.

The Volumetric Time Model does not exist to make you feel small. It exists to make the hidden constraints visible. It exists to separate the clean geometry of what exists from the messy human reality of what can be accessed. It exists to say, yes, the world can be consistent, and no, that does not automatically mean you had usable control. It exists to draw a bright line between being informed and being empowered.

That line matters because modern life is full of prediction, full of dashboards, full of warnings, full of numbers, full of confident language. It is easy to think that more information automatically means more agency. It is easy to confuse visibility with leverage. It is easy to mistake a forecast for a handle you can grab. Then you try to move the outcome and nothing moves, or it moves in a way that does not matter, and you start blaming yourself for failing at a game you were never given the controls for.

This book has been a long argument against that confusion. Not because prediction is bad. Prediction is useful. Prediction is sometimes lifesaving. But prediction without leverage is not power, it is illumination. It is a flashlight in a fog bank. That can help you walk without falling. It can also make you realize how close the cliff is. The flashlight is not the bridge. The flashlight is not the steering wheel. It is a tool that must be paired with something else if you want change.

That something else is reach. A lever. A window. A reversible step. A clear coupling between action and outcome that exists inside the time you still have. When you have that, information becomes fuel. When you do not, information becomes weight.

If you take only one habit from these pages, let it be this habit. When someone shows you a prediction, ask what action comes with it. Ask what can be done now, not later. Ask what changes downstream if you move. Ask how long the window stays open. Ask what reversibility exists if you guess wrong. Ask what it costs to act, in money, energy, reputation, pain, and attention. Ask who pays that cost. Ask who benefits if you wait. Ask who benefits if the undo button is missing.

These questions are not cynicism. They are hygiene. They are what keeps everyday folks from being politely crushed by systems that offer insight without agency. They are what keeps you from being dazzled by a forecast that arrives with no handle attached. They are what turn knowledge into something usable, instead of something that merely proves you were awake while the outcome happened.

There is also a softer ending, and it matters just as much. If you live in a world where delay and noise are real, then a lot of your life is spent managing what you cannot change. That is not defeat. That is navigation. It is skill. It is wisdom in the old sense, not the smug sense. It means choosing what to hold, what to release, what to wait for, and what to refuse to carry. It means learning the difference between a lever and a mirage.

And if you are the kind of reader who likes to argue, here is the part that will irritate you, on purpose. Clarity is not weakness. Simplicity is not childish. A concept that matters should survive contact with everyday language. If it collapses when you remove the fancy wrapping, it was never sturdy. Sometimes the smartest thing a person can do is refuse to be impressed. Refuse to confuse complexity with truth. Refuse to accept a worldview that cannot explain itself without forcing you to feel stupid.

So, imagine the shape again. Imagine the coastline. Imagine the full geometry that exists whether you look at it or not. Then imagine your lived life inside it, a moving point that only receives what can arrive. The gap between those two views is not a spiritual problem. It is an access problem. It is a channel problem. It is a timing problem.

That is good news, in a quiet way. Because timing problems can be redesigned. Channels can be widened. Noise can be reduced. Windows can be made real instead of decorative. Undo can be built into choices so that choices are actually choices. When you demand that kind of design, you are not asking for a perfect future. You are asking for reachable agency.

This is the final promise. You do not need to be able to see everything. You do not need to win metaphysics. You do not need to escape the shape. You just need levers that connect, windows that are honest, and reversibility where it matters. If this book helps you name the difference between what you can see and what you can change, then it has done what it came to do.

The future can be readable while still being unreachable. That is not mysticism. It is timing. The next time you are shown a forecast, do not only ask if it is true. Ask if it is usable. Ask where the lever is. Ask how much time you have. Ask what happens if you pull it. Then decide, not in shame, not in panic, but in the clear daylight of what is actually reachable.

Don't miss out!

Visit the website below and you can sign up to receive emails whenever Ralph Clayton publishes a new book. There's no charge and no obligation.

https://books2read.com/r/B-A-WZAFF-SSTBJ

BOOKS 2 READ

Connecting independent readers to independent writers.

Did you love *The Volumetric Time Model: Why the Future Feels Decided - Deluxe Edition*? Then you should read *The Children of Kings: A Study in Useful Men - Deluxe Edition*[1] by Ralph Clayton!

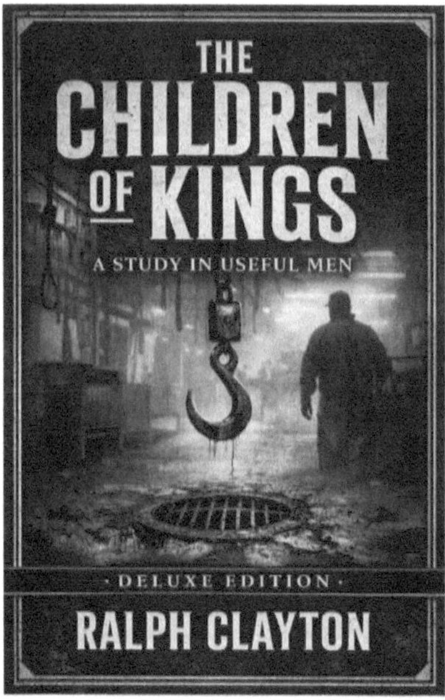

[2]

The Children of Kings: A Study in Useful Men

(The Temporary Arrangements Series — Book 2)

Some men inherit power.Others inherit the damage.

The Children of Kings is the second book in The Temporary Arrangements Series, a dark, unsentimental examination of how men are shaped, renamed, and repurposed inside systems that no longer need to announce

1. https://books2read.com/u/mlVWk7

2. https://books2read.com/u/mlVWk7

themselves with violence. Set against the residue of old hierarchies—money, fear, loyalty, appetite—the book follows the quiet mechanics of recruitment, obedience, and usefulness.

Here, power does not roar.It schedules.It does not destroy bodies. It processes people.

Across industrial backrooms, private compounds, waiting rooms, and administrative corridors, men learn the same lesson in different forms: survival is not freedom, and usefulness is not dignity. Names change. Roles harden. What looks like choice is often just alignment happening slowly enough to feel voluntary.

Violence appears, but it never resolves anything. Tradition lingers, but it no longer decides outcomes. The real work happens in delay, procedure, and the moment when a man realizes he is being maintained rather than protected.

This is not a redemption story.It is not a power fantasy.It is a study.

Written in a stark, noir-inflected style with dark humor and procedural calm, The Children of Kings continues the series' exploration of modern authority, masculinity, and inheritance—where silence carries more weight than spectacle, and nothing is explained away.

Nothing ends.It just keeps processing.

The Deluxe edition includes : Author's commentary and extras.

Read more at https://www.ralphclayton.uk/.

Also by Ralph Clayton

The Spectacle Trilogy - A Precursor Cycle
The Imaginary Pot of Vodka Gold - Final Cut
Tales from the Motherland — Book Zero

The Temporary Arrangements Series
How to Be Nothing by the Majestic Man
The Children of Kings: A Study in Useful Men - Deluxe Edition
Please Remain Seated: This Will Only Take a Moment:
How Hunger Is Measured: A Childhood - Deluxe Edition
You May Now Proceed: Nothing Is Stopping You - Deluxe Edition
You May Now Proceed Nothing Is Stopping You

Standalone
Tales From The Motherland - Deluxe Cut
The Volumetric Time Model: Why the Future Feels Decided - Deluxe Edition

Watch for more at https://www.ralphclayton.uk/.

About the Author

Ralph Clayton writes fiction about power after it stops being dramatic.

His work focuses on systems that do not shout, violence that no longer needs to happen, and lives shaped less by choice than by timing, procedure, and quiet compliance. Across novels and interconnected series, Clayton examines how modern authority removes people politely—through optimization, maintenance, and waiting—rather than spectacle or force.

His books blend dark satire, procedural horror, and existential noir, drawing on post-Soviet realism, institutional absurdity, and contemporary technological anxiety. Recurring themes include exile, erasure, delayed agency, and the slow normalization of the unbearable. Redemption is rare. Resolution is usually administrative.

Clayton's writing is known for its restrained brutality, deadpan humor, and cold clarity. Violence, when it appears, is never heroic. Systems, when exposed, are never personal. Characters survive not by rebellion, but by adaptation—and sometimes by hesitation.

He is the author of *How to Be Nothing*, *The Children of Kings*, *How Hunger Is Measured*, and *Please Remain Seated: This Will Only Take a Moment*, among others. His books are part of a shared narrative universe in which outcomes are fixed, explanations are optional, and continuity is always confirmed.

Ralph Clayton lives quietly and writes regularly.

Read more at https://www.ralphclayton.uk/.

www.ingramcontent.com/pod-product-compliance
Lightning Source LLC
La Vergne TN
LVHW090256210926
848480LV00001B/36
* 9 7 9 8 2 3 3 6 7 4 2 9 7 *